T0253616

Reliability Basic Theories and Applications in Electrical Apparatus

Lu Jianguo, Wang Jingqin, Luo Yanyan, Su Xiuping

Hebei University of Technology
Tianjin, China

iUniverse, Inc.
Bloomington

Reliability Basic Theories and Applications in Electrical Apparatus

iUniverse books may be ordered through booksellers or by contacting:

iUniverse
1663 Liberty Drive
Bloomington, IN 47403
www.iuniverse.com
1-800-Authors (1-800-288-4677)

ISBN: 978-1-4759-5240-7 (sc)
ISBN: 978-1-4759-5242-1 (hc)

Library of Congress Control Number: 2012918478

Printed in the United States of America

iUniverse rev. date: 10/16/2012

PREFACE

Reliability technology is a comprehensive technology that has been developed since 1950s, with probability theory as its theoretical basis and mathematical statistics as its fundamental method. Reliability technology includes reliability design, reliability manufacturing, reliability test, reliability statistics, reliability management and failure analysis, etc. Product reliability is one of the important aspects of product quality. Reliability of the equipment or system is essentially dependent on the reliability of every component used in the system. So, if the reliability of the components used is low, the system reliability can be hardly ensured. In addition, the system reliability will decrease with the increase of the number of components used in the system. As the automatic control system development towards increasing scale, increasingly higher requirements for the component reliability have been posed to ensure normal system work.

The initial reliability research was on electronic components. Some industrial developed countries paid great attentions to the problem of product reliability, and put tremendous manpower and physical resources into reliability research, and made great achievements. In the large-scale projects such as *Apollo* moon flight, space shuttle and space ship, thorough and detailed reliability assurance program has been prepared before hand, which plays a decisive role in the successful completion of these projects, and also reflects the significance of reliability technology.

Electrical apparatus are widely applied in various control systems and power distribution systems, so, it is very important to improve the electrical apparatus reliability for ensuring the normal work of control system and the normal power supply of power distribution system.

The Chapter I in this book is a brief introduction to the history of electrical apparatus development and the content of reliability work; in

Chapter II, we expound the fundamental knowledge on reliability; in Chapter III we expound the basic theory of reliability sampling inspection and design; from Chapter IV to Chapter IX , we expound the reliability indices, reliability test method, reliability verification test sampling plan, test procedure and reliability test device for relays, contactors, miniature circuit breakers, residual current operated circuit breakers, low-voltage circuit breakers and overload relays. The Electrical Apparatus Institute of China Hebei University of Technology has carried out the researches on the electrical apparatus reliability theory and engineering applications since 1978. The most of the content in this book is a summary of the study achievements for over thirty years. We hope it will play a certain role in promoting the further improvement of electrical apparatus reliability.

Allow us to acknowledge Dr. Chi Leung of Metalor Technologies (USA) and Dr. Zhuanke Chen of Chugai USA LLC who have given a considerable help in reviewing and publishing for this book! We sincerely welcome the precious suggestions given from readers.

The authors

5.28.2012

Contents

1

OVERVIEW

The reliability of an electrical product refers to the ability of this product to perform its required functions under stated conditions for a specified period of time. In modern day reliability studies, the advancement of manufacturing must be accompanied by better understanding of the science and technology (S&T) of reliability to solve practical reliability problems.

Reliability technology refers to the engineering methods related to product reliability and has been in continuous development for decades. During the World War II, electronic devices were widely used in military equipment. However, it frequently occurred that various electronic devices were unreliable and could not be effectively put into service. In a battle in the early 1950s, this problem was so remarkable that the radar equipments of the U.S. were frequently incapable of normal operation and idle in repair-waiting state. The high repair and maintenance cost caused by low reliability of electronic devices lead the U.S. industry to start to attach importance to the reliability problems. This embarked the investigation, research and test works that raised the curtain of reliability research in the field of electronic devices. The early researches were focused on the electron tubes and their performance. This focus was not only on its electrical properties, but also attached importance to its resistance to shock and impact and other environmental adaptability.

The period from the 1950s to 1960s was the decades of rapid development of reliability technology during which the U.S. Defense Department established various reliability research organizations, such as: the AdHoC reliability group in 1950, the AGREE(Advisory Group on Reliability of Electronic Equipment) in 1952, and the ACGMR (AdHoC Group on Missile Reliability) in 1957, etc. Led by these organizations, the works of reliability management, analysis and test, etc. were carried out in an extensive scale. And the reliability researches were also developed extensively by various related technical associations, companies and manufacturers in the U.S. In the later period of 1960s, many standards on reliability management, design and test appraisal, etc. were published, and the reliability technology on electronic elements and devices became increasingly mature. In the 1970s, the reliability research in the U.S. has gradually penetrated into the fields of mechanics, electrical engineering, electrical power and chemical industry, etc.

Overall, the U.S. developed the earliest, most extensive and effective researches on reliability. In addition, Japan, Britain, France, Germany and the former Soviet Union, etc. have also developed the reliability researches and achieved tremendous result. The former Soviet Union has not only prepared many fundamental reliability standards, but also prepared the reliability standards and reliability requirements for many products. They also specified reliability test methods in the product standards, and published many books and manuals on reliability. For instance, the Technical Manual on System Reliability that was published in 1985, compiled by И.А.ушаков.

1 SURVEY OF ELECTRICAL APPARATUS RELIABILITY

1.1 SURVEY OF ELECTRICAL APPARATUS RELIABILITY IN INDUSTRIALLY DEVELOPED COUNTRIES

Electrical apparatus manufacturers and research institutes in the industrial developed countries have identified the research and application of electrical apparatus reliability as a major task. Even in those developed countries, the stipulation of reliability indices is not uniformly applied. The result is that some electrical apparatus have and some do not have properly stipulated reliability indices. Product reliability has been controlled in most factories, and the level of product reliability has become the major competition among enterprises. The current research and application of electrical apparatus reliability in industrial developed countries are mainly focused on the following aspects:

1.1.1 FORMULATION OF RELIABILITY STANDARDS

The formulation of reliability standards has developed very fast, especially the preparation of fundamental reliability standards. Since established in 1965, the

IEC TC56 Technical Committee for Reliability has published many standards on reliability and maintainability. In 1988, IEC TC56 held an annual conference in Tokyo and determined to constitute the TC56 standard system by adopting the "toolbox principle". The so called "toolbox principle" means dividing the standards and documents into four categories:

1. Top-level document (IEC300) - Management on Reliability and Maintainability is a fundamental standard on reliability management.
2. Application guide - mainly include reliability specification, reliability design analysis, reliability prediction of elements (components), reliability test, reliability growth, reliability screening, software reliability, maintainability technology and field evaluation.
3. Tool document - mainly include two system standards of IEC605 Equipment Reliability Test and IEC706 Maintenance Guidance.
4. Support document.

The evolution of reliability standards for electrical apparatus can be captured with the following milestones:

- The first reliability standard for electrical apparatus with reliability requirements was published in 1964 as the U.S. Military Standard MIL-R-39016 General Specification on Electromagnetic Relay with Reliability Index.
- Japan has issued several standards afterwards: 1980 the Japanese industrial standard JIS C5440 *General Reliability Rules of Small Relay used in Control Circuits,* 1981 the test method of failure rate both in JIS C4530 *Clapper Type Electromagnetic Relays* and 1982 the JIS C4531 *Contactor Type Relays.*
- In 2002, 10 terms about reliability were listed in IEC60050-444 *International Electrotechnical Vocabulary: Elementary Relays.*
- In 2011, IEC issued the IEC 61810-2-1 Electromechanical Elementary Relay-Part 2-1: Reliability.
- 1983, the former Soviet Union has listed the content of reliability requirement and test in many standards for electrical apparatus, such as the ГOCT 12434-83 *General technical conditions for low voltage switch electrical appliance* which stipulated the reliability requirement of product.
- Germany specified the ultimate on/off times to be achieved by the contactor with rated value of product mechanical life and electrical life based on the

performance of 90% of all contactors in the VDE0660 *"Specification of Low-voltage Switch Electrical apparatus*. This became the means to the concept of 0.9 reliability in the mechanical life and electrical life of contactors.

- In France NFC 63–100 Industrial Low-voltage Control Equipment: Contactor, it's specified that for the batch produced electrical apparatus, especially those with agreed thermal current of less than or equal to 40A, the mechanical life is tested on representative prototype by repeat mode, and the manufacturer should provide the mechanical life value of the product after statistics of test results, which actually means to determine the mechanical life of contactors by the concept of reliability.

1.1.2   RELIABILITY TEST AND DEVELOPMENT OF RELIABILITY TEST DEVICE

Computers are commonly adopted for control and detection in the reliability life test of electrical apparatus in the U.S. and Japan.

For example, the Japanese Yaskawa Company adopts the auto test device that uses electronic computer for control and detection in the contact reliability test of relays; Panasonic Company adopts the microcomputer-controlled full auto test device in the reliability life test of relays; Fujitsu Company also adopts the computer-controlled test device in the reliability test of reed relays, which is capable for the life test of 200 reed pipes at the same time.

In the U.S., the reliability life tests of RT160 relays are with microprocessor-controlled systems that come with the functions of auto measurement of contact pressure drop at contact point and other parameters, display and print of test results, etc.

German electrical apparatus companies frequently conduct life tests of contactors. For example, the contactors produced by Siemens are sampled each week for reliability testing. About 20~30 sets for the low current contactors ($I_N \le$ 32A) and 10~15 sets for the higher current contactors (I_N >32A) are sampled every two weeks as a group for mechanical life test until all test samples failed. The reliable life at reliability R=0.9 is determined by Weibull probability Paper according to the test result, which should not be less than the specified mechanical life value of the product sample. As for Siemens contactors, the number of sets of each group is the same for electrical life test and mechanical life test, but the tests are conducted once by sampling each month, and the products sampled each year constitute one group.

The contactors produced by French Telemecanique Company are sampled 10 sets each month for mechanical life test. The Japanese S-type contactors are also sampled 2~20 sets each month for mechanical life test.

1.1.3 RESEARCH ON ACCELERATED LIFE TEST

The mode and method of accelerated life test and data analysis method are researched in the U.S., Japan and many other countries. For example, in Japan, the accelerated life test of switches is conducted by using load voltage and current as accelerating variable; additionally, the accelerating factor at different load voltage and current values are calculated according to the test results.

1.1.4 RESEARCH ON RELIABILITY DESIGN

The U.S., Japan and many other countries attach great importance to the reliability design of products. For example, the leading electrical apparatus companies of the U.S., Japan and Germany designed and developed intelligent circuit breaker, which significantly improved the reliability of power supply. Advanced intelligent circuit breaker can not only transmit the signal to the computer in control room from remote distance, but also receive computer instructions, to realize system automation and two-way communication.

1.1.5 RESEARCH ON RELIABILITY PHYSICS

The reliability physics has been researched since 1960s in foreign countries. The U.S. Air Force ROME Air Development Center started the on-site failure analysis in 1960s of the failed element. J. Vaccro first proposed using the concept of "failure physics" for the research on element reliability. The conference on "failure physics" has been held every year since 1962 in the U.S and in 1967 renamed to conference on "reliability physics". The so called "reliability physics" is a special science on the product failure mechanism. It studies the specific physical and chemical process on the way and reason of product failure. It can be seen that the research on reliability physics is a fundamental research to improve product reliability.

Great importance was placed on the reliability physics in the U.S., Japan and other countries. In the area of electromagnetic relays and contactors reliability, in-depth research of the performances of electric contact, electromagnetic system and other parts of the contactor are conducted. The various forms and causes of the faults of operating relay are mastered through large amount of tests and investigation in actual application. Among others, special importance is placed on the contact reliability of the electric contact. They studied the influence of various application conditions on the contact reliability, conducted overall analysis on the material, shape, contact mode, contact pressure, etc. of the contact point, and proposed requirements for the product design and application. In electromagnetic contactor, a great deal of investigations have been conducted into the faults of operating contactors in Japan, and it's shown by the results that the reliability of contactor is not only determined by its design and manufacturing, but also significantly influenced by correct or wrong application.

1.2 SURVEY OF CHINESE ELECTRICAL APPARATUS RELIABILITY

The development history on reliability of Chinese electrical apparatus can be divided into the following several periods:

1.2.1 PERIOD OF POPULARIZING RELIABILITY KNOWLEDGE AND RESEARCH

The former Ministry of Machine-Building Industry paid great attention to the reliability of electrical products. As early as the late 1970s, it entrusted Hebei University of Technology to open classes to train the latest technology on electrical products and established that reliability technology be one of the main education contents. In October 1983, China Electrotechnical Society formed the Electrical Product Reliability Research Committee to further promote reliability research on electrical products. Since then, all academic communication activities were developed and organized under the leadership of the committee. Electrical product reliability training classes have since been organized many times to promote the importance of reliability development of Chinese electrical apparatus industry.

1.2.2 PERIOD OF PROMOTION BY THE GOVERNMENT

The former Ministry of Machine-Building Industry had convened several reliability working conferences. In 1986, it announced the "Improvement Of Electrical Product Reliability Notice" and assigned work to "Review The Reliability Index Of Mechanical And Electrical Products Within Limited Time" in mechanical and electrical industry. A total of 7 document batches (1,189 specifications in total) on mechanical and electrical products were issued with reliability index to be reviewed within limited time in 1986~1991. This industry wide review included manufactured products such as relays, contactors, transformers, measuring relays, motors, power electronic devices and many other electromechanical products. The result of the review played a significant role in promoting the development of reliability of Chinese mechanical and electrical products. More than twenty national standards on reliability have now been established in China.

1.2.3 PERIOD OF SETTING STANDARDS, TEST METHODS, TEST EQUIPMENTS AND IMPLEMENTATION IN INDUSTRY

1.2.3.1 FORMULATION OF STANDARD ON ELECTRICAL APPARATUS RELIABILITY

Hebei University of Technology and other institutions published five electrical apparatus reliability standards which were approved and issued by the National Development and Reform Commission and in 2007. The standards were adopted in 2008 by Standardization Administration of the People's Republic of China SAC as national standards 2008. By December 2008, eight national standards on the reliability of relays and representative low-voltage electrical apparatus had been issued in China:

- GB/T 15510-2008 "General Rules for Reliability Test of Electromagnetic Relay for Control Circuits";
- GB/Z 10962-2008 "General Rules of Reliability for Machine Tool Electrical Components";

- GB/Z 22200-2008"Reliability Test Method for Lower Capacity Alternating Current Contactors";
- GB/Z 22201-2008"Reliability Test Method for Contractor Relay";
- GB/Z 22202-2008 "Reliability Test Method of Residual Current Operated Circuit-breaker for Household and Similar Uses";
- GB/Z 22203-2008"Reliability Test Method of Over-current Protection Circuit-breakers for Household and Similar Installation";
- GB/Z 22204-2008"Reliability Test Method for Overload Relay";
- GB/Z 22074-2008 "Reliability Test Method for Molded Case Circuit-breakers".

These standards have regulated the reliability index, reliability test method, failure criteria and reliability verification test scheme of relay and representative low-voltage electrical apparatus.

1.2.3.2   RESEARCH ON RELIABILITY TEST TECHNOLOGY AND DEVELOPMENT OF DETECTION DEVICE

Hebei University of Technology established an electrical apparatus reliability test center and has passed the approval of CNACL in May 2001 to service the industry. It can certify the electrical apparatus reliability of products, and has developed reliability test equipment which conforms to the above national standards. Some of the equipment in the center are as shown: Figure 1-1, overview of relay reliability test device; Figure 1-2, overview of contactor reliability test device; Figure 1-3, overview of miniature circuit-breaker reliability test device; Figure 1-4, overview of residual current operated circuit-breaker reliability test device; Figure 1-5, overview of overload relay reliability test device; and Figure 1-6, overview of molded case circuit-breaker reliability test device.

Figure 1-1 Overview of relay reliability test device

Figure 1-2 Overview of contactor reliability test device

Figure 1-3 Overview of miniature circuit-breaker reliability test device

Figure 1-4 Overview of residual current operated circuit-breaker reliability test device

Figure 1-5 Overview of overload relay reliability test device

Figure 1-6 Overview of molded case circuit-breaker reliability test device

1.2.3.3 ESTABLISHMENT OF RELIABILITY LABORATORY IN CHINESE ELECTRICAL APPARATUS INDUSTRY

To implement the China national standards on the reliability of electrical apparatus listed above, Hebei University of Technology, in cooperation with Yueqing government (important production base of Chinese electrical apparatus) established the Reliability Laboratory of Yueqing City and Hebei University of Technology (see Figure 1-7). Hebei University of Technology also helped more than ten electrical apparatus manufacturers to establish electrical apparatus reliability test laboratories (see Figure1-8~Figure1-11). These included China Electrical Apparatus Product Quality Supervision and Inspection Center, Changshu Switchgear Mfg. Co., Ltd., Xiamen Hongfa Electroacoustic Co., Ltd., and PEOPLE ELECTRICAL APPARATUS GROUP CHINA. These electrical apparatus manufacturing leaders have developed product reliability tests and failure analysis; favorable results have been obtained for improving their product reliability.

Figure 1-7   Reliability Laboratory of Yueqing City and Hebei University of Technology

Figure 1-8   Electrical Apparatus Reliability Laboratory in China Electrical Apparatus Product Quality Supervision and Inspection Center

Figure 1-9 Relay reliability laboratory in Xiamen Hongfa Electroacoustic Co., Ltd.

Figure 1-10 Electrical apparatus reliability laboratory in Changshu Switchgear Mfg. Co. Ltd.

Figure 1-11 Reliability laboratory in People Electrical Apparatus Group China

1.2.4 PERIOD OF COMBINATION OF PRODUCT RELIABILITY & ELECTRICAL CONTACT RESEARCH WITH INTERNATIONAL CONFERENCES

The reliability of electrical products largely depends on the condition whether it encounters faults frequently during the application. As a result, the reliability can be improved by making improvements in design, manufacturing and materials after carrying out failure analysis for the electrical products to find fault mode and mechanism. Electrical contact malfunctions (including high contact resistance and contact welding) are usually one of the major failure modes of different electrical apparatus.

From 1974 to 1977, Circuit-breaker Reliability Group of IEEE Switch Device Technical Committee investigated the faults of high-voltage circuit breakers used worldwide and 77,892 circuit breakers from 102 companies in 22 countries were investigated. These circuit breakers are mainly in the range of rated voltage at or above 63kV and they have been used in the field since 1964. It was found that 19% operation faults of circuit-breaker were from electrical contact failure in auxiliary circuit and control circuit.

Electrical contact failure accounts for 75% among all the relay faults. Therefore, studying the electrical contact is closely related to the reliability of electrical products. In China, there are eight academic annual meetings on electrical contacts. Recently it has been combined with the conference of reliability that is organized by the Electrical Product Reliability Research Committee of China Electrotechnical Society and it is named as ICREPEC "International Conference on the Reliability of Electrical Products and Electrical Contacts". So far three ICRPEC conferences have been held in August 2004, March 2007 and September 2009 respectively. The 4th conference was the joint meeting of ICREPEC and the 26th International Conference of Electrical Contact (ICEC) in Beijing, China in May 2012.

1.2.5   PERIOD OF EVALUATION ON THE RELIABILITY OF ELECTRICAL APPARATUS

The eight China national standards published in 2008 on electrical apparatus reliability discussed above only regulate the reliability test method but do not regulate the reliability evaluation. This led to the establishment in January 9, 2012 of *Electrical apparatus Reliability Promotion Committee* in China. Under the guidance of this new committee, the Electrical Product Reliability Research Committee of China Electrotechnical Society, Reliability Committee of China Association for Standardization, Hebei University of Technology and the representatives of electrical apparatus manufacturers in China formulated a series of association standards for the reliability evaluation of relays and a series of low-voltage electrical apparatus. These new standards will play a positive role in improving the reliability of Chinese electrical apparatus.

The electrical apparatus reliability evaluation standards under formulation are as follows:

- CAS 213-2012   General Principles of the Reliability Evaluation of Relays, Controlling and Protective Electrical Apparatus
- CAS 214-2012   Evaluation on the Reliability of Molded Case Circuit-Breakers

- CAS 215-2012   Evaluation on the Reliability of Control Electromagnetic Circuit- breaker
- CAS 216-2012   Evaluation on the Reliability of Overcurrent Protection Circuit- breaker in Household and Similar Places
- CAS 217-2012   Evaluation on the Reliability of Residual Current Operated Circuit- breaker in Household and Similar Application
- CAS 218-2012   Evaluation on the Reliability of AC Contactor
- CAS 219-2012   Evaluation on the Reliability of Overload Relay

The development of electrical apparatus reliability evaluation above indicates that reliability of Chinese electrical products enters a new phase.

1.3 PROSPECT ON RELIABILITY OF CHINESE ELECTRICAL APPARATUS

Many researches have been done in China on the reliability of Chinese electrical apparatus to understand and to improve their reliability, and a lot of progresses have been made. However, since the reliability study in China started quite late, the design and manufacturing impact on reliability have not been seriously considered in past development of new products; and many manufacturers do not emphasis enough on reliability. Therefore, more work is needed to educate and emphasize about reliability in order to improve the reliability of Chinese electrical apparatus.

At present, an additional series of reliability evaluation standards of Chinese electrical apparatus are under approval after first draft formulation, consultation and experts' examination. *Electrical apparatus Reliability Promotion Committee* has decided to develop some demonstration projects of electrical apparatus reliability in Chinese key electrical apparatus enterprises upon the approval and issuance of these standards. It is believed that the reliability of Chinese electrical apparatus will be improved significantly under the leadership of *Electrical* apparatus *Reliability Promotion Committee,* the organization of Electrical Product Reliability Research Committee of China Electrotechnical Society and the active participation of Chinese key electrical apparatus enterprises.

2   DEFINITION OF RELIABILITY

The reliability of a product refers to the ability of this product to perform its required functions under stated conditions for a specified period of time (or by specified operating cycles)

First, product reliability is related to the specified functions. The so called required functions refer to the various technical performances specified in the product standard or specification. "To perform its required functions" mentioned in the above definition means to perform all of the specified technical performances.

Second, the product reliability cannot be separated from the specified conditions. The so called specified conditions refer to the load conditions, environmental conditions and storage conditions during product application. It's obvious that the product comes with different reliability in different load conditions. For example, the electrical apparatus reliability is influenced by the amount of its contact on/off current and contact loop power voltage, as well as the environmental conditions (such as temperature, humidity, height above sea level, salt fog, impact and vibration, etc.) It's obvious that the electrical apparatus will have lower reliability in severe environmental conditions. The electrical apparatus reliability is also influenced by storage conditions; e.g., the reliability of products would be lowered under poor storage conditions and the products would become damp or damaged as well.

Finally, the close relationship between product reliability and specified time is important. It's obviously easier for the product to complete the specified functions for a day than to complete the same specified functions for a year. In other words, the longer the specified time, the lower is the product reliability, i.e. the product reliability is lowered with the increase of its application time.

The above definition of reliability can only qualitatively describe the level of product reliability. To describe the level of product reliability quantitatively, the definition of confidence is introduced as follows. The electrical apparatus confidence refers to the probability of the product to perform the required functions in specified conditions for a specified time, generally expressed in R. For example, suppose the confidence of a contactor of certain specification operated for 10^6 times is 90%, that means when n contactors of the same specification are sampled for several times and operated for 10^6 times in specified conditions, 90% of the contactors on average can perform the required functions at the specified conditions.

3 RELATION BETWEEN PRODUCT RELIABILITY AND QUALITY

Product reliability is an importance aspect of product quality. The quality of electrical apparatus should include its technical performance index and reliability index, which are related to and different from each other. If the product has low reliability, even though it has advanced technical performance, it cannot be considered to be a good quality product.

For example, suppose a low-voltage circuit breaker has advanced on/off capacity index, but unreliable action, it cannot reliably act at short circuit or other fault, and the accident may consequently be expanded. So it is obvious that this low-voltage circuit breaker cannot be considered as a product with good quality. On the contrary, suppose a low-voltage circuit breaker has very low on/off capacity but more advanced reliability, it can reliably act at short circuit, but it can only be applied in the place with low short-circuit current, so it's obvious that this low-voltage circuit breaker also cannot be considered as a product with good quality because of its restricted application conditions. Therefore, as to a product of high quality, both high reliability and advanced technical performance index are indispensable.

4 INHERENT RELIABILITY AND USE RELIABILITY

It's stated in IEC300 that: "the reliability displayed by the product in user's hands is the reliability that is the most meaningful for users." The reliability of the product displayed during actual application by the user is called Operational Reliability which is composed of Inherent Reliability and Use Reliability. The inherent reliability refers to the product reliability determined by the manufacturer during the process of production, closely related to the selection, design, manufacturing, test of raw material and part/component, and other factors. It's the reliability that is determined in the standard environment simulating actual operational conditions and must be ensured by the manufacturer. The use reliability is the reliability determined by some man-made factors related to the product application, such as the operating situation, maintenance method and technique after the product is manufactured. These man-made factors continuously influence the product during actual application, and all of them may significantly influence the reliability of the product.

If a product with high inherent reliability is used with low use reliability because of the improper application, then the operational reliability of the product is undesirable. On the contrary, if a product with low inherent reliability is used in a high use reliability way because of its proper application, the operational reliability of the product is still undesirable, but can meet certain requirements.

5 FAILURE LAWS

The failure rate of the product $\lambda(t)$ refers to the probability of failures occurred in unit time after the moment t when the product has operated until the moment t. The relation curve between the failure rate $\lambda(t)$ of many products and time t is shown in Figure 1-12, in which, the curve is usually called "tub curve". It can be seen from the curve that the change of product failure rate with time can be approximately divided into three periods:

1. Early Failure Period: it occurs in the early period of product operation. It's featured by higher product failure rate that decreases with the increase of operating time. The cause of product failure in this early period is the defective design and manufacturing technology, such as defective raw material, poor manufacturing technology, poor environmental sanitation of production, production device fault, negligence of operating personnel, and careless quality control, etc.

2. Accidental Failure Period: the product failure is random in this period. Its characteristic is that: the failure rate of the product is low but stable, and it is close to a constant, this period is the best operating period of the product.

3. Loss Failure Period: it occurs in the later period of product operation. Its characteristic is that: the failure rate of the product significantly increases with the increase of operating time. The product failures in this period are mainly caused by aging, wearing and fatigue, etc.

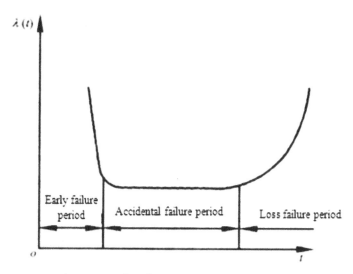

Figure 1-12   Product Typical Failure Rate Curve

6   SIGNIFICANCE OF IMPROVING ELECTRICAL APPARATUS RELIABILITY

The improvement of electrical product reliability is a significant aspect in the improvement of electrical apparatus quality. Many various types of electrical apparatus are widely used in various sectors of our national economy and play an important role in it too. However, the current quality of various electrical apparatus made in China is still not very desirable, and their faults usually lead to abnormal operation of various other systems, and consequently cause tremendous economic loss. With the development of more S&T and industrial production, the scale of automatic control system is increasing, and a large-scale automatic control system usually uses tens or even hundreds of thousands of elements, and the system reliability is closely related to the number of elements. Suppose the system is a system in series connection of reliability (i.e. merely failure of one element in the whole system will lead to system failure), the system confidence R_s equals to the product of the confidences of all elements in the system, i.e.

$$R_s = \prod_{i=1}^{n} R_i \qquad\qquad (1\text{-}1)$$

Where: n is the quantity of elements used in the system; Ri is confidence of every element (i=1,2,...,n).

Suppose the confidence of all elements used in the system in series connection of reliability is 0.99999, when *n* is a different value, the system confidence value can be obtained by equation (1-1), as shown in Table 1-1.

Table 1-1 Relation between System Reliability R_s and Quantity *n* of its Used Elements

n	100	1,000	10,000	100,000
Rs	0.999	0.99	0.905	0.368

It can be seen from Tale 1-1 that the system confidence rapidly decreases with the increase of the quantity of elements used in the system. Suppose to ensure a system confidence of 0.95 when $n=10^4$, the confidence of every element must reach 0.9999949. It can be seen that the larger system, the higher requirement for the reliability of its used elements.

In summary, to improve the electrical apparatus reliability is the requirement of the development of national economy, and it has great significance.

7 BASIC CONTENT OF RELIABILITY WORK

There are many factors that can influence the product reliability. The various stages from the determination of product reliability index, research, design, manufacturing, tests and appraisals, to putting the product into service can affect the product reliability. The failure analysis after product failure is also closely related to the product reliability. For electrical apparatus, it requires a lot of time and cost in reliability testing, the failure analysis of the failed product is of particular importance to learn the reasons of failure.

Figure 1-13 shows the block diagram of the basic contents of product reliability works.   It illustrates the whole process flow from design to manufacture to reliability screening. This leads to more in depth reliability test results and in parallel with knowledge from field applications data. The results of product failure analysis are feedback to the design phase with another cycle of study and improvements.

It should be noticed that the product reliability is also closely related to the reliability management, to improve product reliability, we must have excellent reliability organization and seriously conduct reliability management.

Finally, it should also be pointed out that the product reliability design, determination of sampling scheme at reliability test, and the evaluation of test result involve many mathematics works such as Boolean algebra, probability theory and other mathematic tools, i.e. the so-called reliability mathematics. This means that the reliability technology is also closely related to the science of reliability mathematics.

In subsequent chapters in this book, we will explain the reliability sampling theory, reliability design theory of electrical apparatus with explanations of the basic knowledge of reliability, and finally discussed the reliability of electrical products such as relay, contactor, miniature circuit breaker, residual current operated circuit-breaker, low-voltage circuit breaker, overload relay and other electrical apparatus.

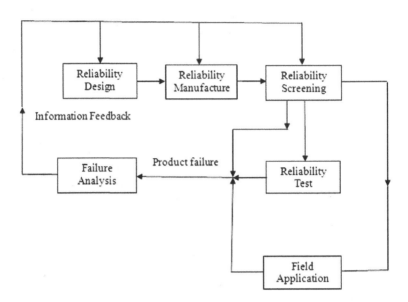

Figure 1-13 Block Diagram of Basic Content of Reliability Work

2

BASIC KNOWLEDGE OF RELIABILITY

1 BASIC KNOWLEDGE OF BOOLEAN ALGEBRA

1.1 BASIC RELATIONS OF BOOLEAN ALGEBRA

Boolean algebra refers to the relation f $(x_1, x_2, x_3...)$ achieved through union operations (U), intersection operations (∩)and negation operations (‾)of Boolean variables $x_1, x_2, x_{3...,}$ Herein, the values of Boolean variables $x_1, x_2, x_3...$can only be 0 or 1 rather than other values. Therefore, the function value of Boolean algebra can also only be 0 or 1.

      Basic relations of Boolean algebra usually refer to the following relations between Boolean variables:

1) Commutative law	x1+x2= x1+x2	(2-1)
	x1x2= x1x2	(2-2)
2) Associative law	x1+(x2+ x3)= (x1+x2)+ x3	(2-3)
	x1(x2x3)= x1(x2x3)	(2-4)
3) Absorption law	(x1+x2) x3= x1	(2-5)
	x1+x1 x2= x1	(2-6)
4) Distribution law	x1 (x2+ x3)= x1x2+ x1 x3	(2-7)
	x1+x2x3= (x1+x2)(x1+x3)	(2-8)
5) Idempotent law	x1+x1= x1	(2-9)
	x1x1= x1	(2-10)
6) Complementarity	$x_1 + \overline{x}_1 = 1$	(2-11)
	$x_1\overline{x}_1 = 0$	(2-12)

7) De Morgan theorem $\overline{x_1 + x_2} = \overline{x}_1 \overline{x}_2$ (2-13)

                                        $\overline{x_1 x_2} = \overline{x}_1 + \overline{x}_2$ (2-14)

1.2 EXPANSION THEOREM

Let $y = f(x_1、x_2、x_3...)$, The above Boolean function is f_1 when making $x_i=1$; The above Boolean function is f_0 when $x_i=0$, then for any value of Boolean variable x_i, the Boolean function y can be expanded into:

$$y = f_1 x_i + f_0 \overline{x}_i$$ (2-15)

This theorem is called addition-shaped expansion theorem.

2   FAILURE DENSITY FUNCTION AND CUMULATIVE FAILURE DISTRIBUTION FUNCTION

When a batch of products are performing life test, failure time (for the electrical apparatus which are operated frequently, it refers to the number of operations until the components failure is detected in the test.) of each product may vary a lot, but the failure data follow certain laws. In the words of mathematical statistics, failure time of products obeys certain distributions. If we handle the failure data properly, we will find the laws which reflect the essence of matters. Drawing failure frequency histogram and cumulative failure frequency histogram is a kind of processing method for the failure data. We can directly find out the general distribution of failure data from the failure frequency histogram.

2.1 FAILURE FREQUENCY HISTOGRAM

Randomly select n products from a batch of products and perform the life test. Suppose the failure time (i.e. service life) of each test sample is $t_1, t_2, ..., t_n$, then we can handle in the following ways.

(1) Divide the failure time into several ranges with a certain time interval Δt (Δt is called the class interval), that is, to divide n failure data into several groups. The group number k can be determined by the formula (2-16).

$$k = 1 + 3.3 \lg n$$ (2-16)

(2) Make a table to represent the range of failure time and values of t_{zi}, Δm_i, f_i^*, F_i, f_i of each group. Herein t_{zi} is the mid-value of failure time range of group i; Δm_i is the number of failure data in group I; f_i^* is the failure frequency of group i, that is, the ratio between the frequency Δm_i and the total number of failure data of group i, i.e.

$$f_i^* = \frac{\Delta m_i}{n} \qquad (2\text{-}17)$$

F_i is the cumulative failure frequency of group i, that is, the sum of failure frequency from group one to group i, i.e.

$$F_i = \sum_{j=1}^{i} f_j^* = \sum_{j=1}^{i} \frac{\Delta m_j}{n} \qquad (2\text{-}18)$$

f_i is the ratio between the failure frequency f_i^* of group i and the class interval Δt, i.e.

$$f_i = \frac{f_i^*}{\Delta t} \qquad (2\text{-}19)$$

(3) Draw the failure frequency histogram and take the failure time t as abscissa and f_i as ordinate. The histogram made in the form of rectangles is called failure frequency histogram.

2.2 FAILURE DENSITY FUNCTION

If we shorten the group interval Δt during data processing (i.e. divide more groups), a new failure frequency histogram can be achieved. The distribution situation coincides with Figure 2.1, but altitude differences of adjacent rectangle are much smaller. When there are more and more test data, and the class interval is shortened continually, i.e. when Δt is smaller and smaller, contour lines on the top of each rectangle in failure frequency histogram will be close to a smooth curve, it is just the failure density curve, as shown in Figure 2-2. Its mathematical expression $f(t)$ is called failure density function.

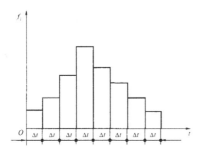

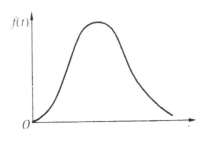

Figure 2-1

Failure frequency histogram

Figure 2-2

Failure density function curve

Because of the sum of areas of each rectangle in Figure 2-1 is equal to 1, so the area between the failure density curve and the abscissa axis in Figure 2-2 is also equal to 1. It's a very important feature of failure density function $f(t)$, i.e.

$$\int_0^\infty f(t)dt = 1 \tag{2-20}$$

The probability $P(a \leq L \leq b)$ that failure time L values between the interval $[a, b]$ is equal to $\int_a^b f(t)dt$, i.e.

$$P(a \leq L \leq b) = \int_a^b f(t)dt \tag{2-21}$$

2.3 CUMULATIVE FAILURE FREQUENCY HISTOGRAM

Take the failure time t as abscissa and the cumulative failure frequency F_i as ordinate, and draw the histogram, achieve the cumulative failure frequency histogram, as shown in Figure 2-3.

2.4 CUMULATIVE FAILURE DISTRIBUTION FUNCTION

If there are many test data, the class interval Δt is shortened continually(i.e. the number of groups is larger and larger), the connection of vertexes at the top right corner of each rectangle in cumulative failure frequency histogram will be close to a smooth curve. This curve is called the cumulative failure density curve, as shown in Figure 2-4. Its mathematical expression $F(t)$ is called cumulative failure distribution function.

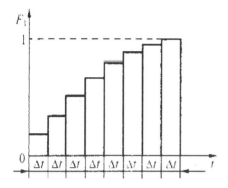

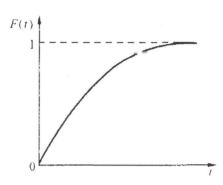

Figure 2-3 Cumulative failure frequency histogram

Figure 2-4 Cumulative failure frequency distribution curve

3 CHARACTERISTIC PARAMETERS OF ELECTRICAL APPARATUS RELIABILITY

Electrical apparatus can be divided into two major groups: repairable products and non-repairable products. Non-repairable products refer to the products which cannot be repaired or can be repaired but not worth repairing. Some small electrical apparatus such as small-capacity AC contactors and small size intermediate relay can be considered as non-repairable products. Repairable products refer to the products which can be repaired. Some large scale electrical products such as low-voltage circuit breakers and high voltage circuit breakers can be considered as repairable products.

3.1 RELIABILITY CHARACTERISTIC PARAMETERS OF NON-REPAIRABLE ELECTRICAL APPARATUS

3.1.1 RELIABILITY

Reliability refers to the probability of performing specified functions under stated conditions for the specified period of time (or number of operations).Reliability R

is the function of time t which is usually represented by $R(t)$. It's called the reliability function. The probability can be used to represent the reliability function; it refers to the probability that the random variable of product life L is not less than the specified period of time t, i.e.

$$R(t) = \begin{cases} P(L \geq t) & (t \geq 0) \\ 1 & (t < 0) \end{cases} \qquad (2\text{-}22)$$

Approximate expression: Select n products and perform the test, if there are total $m(t)$ products failure within the specified period of time t (failure means the products which had lost its specified functions), then the reliability of the product is approximately equal to

$$R(t) \approx \frac{n - m(t)}{n} \qquad (2\text{-}23)$$

The value range is 0-1, i.e. $0 < R(t) \leq 1$.

3.1.2 CUMULATIVE FAILURE PROBABILITY

Cumulative failure probability refers to the probability of losing the specified functions under stated conditions for a specified period of time (or number of operations).Cumulative failure probability is a function of the time t, sometimes it's called "unreliable function". It is represented by $F(t)$. It refers to the probability that the random variable of product life L is less than the specified period of time t, i.e.

$$F(t) = \begin{cases} P(L < t) & (t \geq 0) \\ 0 & (t < 0) \end{cases} \qquad (2\text{-}24)$$

Approximate expression: If there are n products to perform life test and the number of failure at the moment t is $m(t)$, when n is large enough, the cumulative failure probability of the product at the moment t is approximately equal to

$$F(t) \approx \frac{m(t)}{n} \qquad (2\text{-}25)$$

Cumulative failure probability $F(t)$ is a non-decreasing function of the time t. The value range is $0 < F(t) \leq 1$.

3.1.3 FAILURE RATE

The failure rate of the product refers to the probability of failures occurred in unit time after the moment t when the product has operated until the moment t. Failure rate is the function of time t. It's usually called the failure rate function and represented by $\lambda(t)$.

Approximate expression: Premise n products begin to work at the moment $t = 0$, the number of failure at the moment t is $m(t)$ while the number of failure at the moment $t + \Delta t$ is $m(t + \Delta t)$, then the failure rate $\lambda(t)$ can be calculated through the following formula:

$$\lambda(t) = \frac{m(t + \Delta t) - m(t)}{[n - m(t)]\Delta t} \qquad (2\text{-}26)$$

Unit of failure rate: h^{-1}, $10^{-5}h^{-1}(\%/10^{3}h)$, 1/time, 1/10times, $\%/10^{4}$times.

3.1.4 AVERAGE LIFE

The average life of non-repairable products refers to the mean working time (or mean number of operations) of products before failure. It's usually represented as MTTF (Mean Time to Failure).

When we need to know the average life of a batch of products, we should consider the lives of all products in the batch of products as a population. Because of the life test is destructive, so we usually randomly select n products and perform the life test, and their lives tested are t_1, t_2, ..., t_n, this group of life data constitute a sample. The average life $\bar{t}$ of the sample can be ascertained through the formula (2-27):

$$\bar{t} = \frac{1}{n} \sum_{i=1}^{n} t_i \qquad (2\text{-}27)$$

If n is very large, then the calculation through the formula (2-27) will be too cumbersome. So we can divide n life data into k groups according to a certain time interval (or the number of operations) and take the medium value of failure time ranges of each group as the approximate value of each life data in this group. Now the average life of the sample can be calculated through the formula (2-31) and (2-32):

$$\bar{t} = \frac{1}{n} \sum_{i=1}^{k} t_{zi} \Delta m_i \qquad (2\text{-}28)$$

where, Δm_i = the frequency of group i

t_{zi} = the medium value of failure time range of group i

3.1.5 STANDARD DEVIATION OF LIFE

The characteristic quantity of reliability which reflects the discrete degree of product lives is the standard deviation of lives. Premise the sample of a batch of products lives are t_1, t_2, ..., t_n, and the average life of the sample is $\bar{t}$, then the life variance is

$$s^2 = \frac{1}{n-1} \sum_{i=1}^{n} (t_i - \bar{t})^2 \qquad (2\text{-}29)$$

The standard deviation of lives of the sample is

$$s = \sqrt{\frac{1}{n-1} \sum_{i=1}^{n} (t_i - \bar{t})^2} \qquad (2\text{-}30)$$

If the sample is very large (i.e. n is large), the calculation of the life variance s^2 and the standard deviation of lives s of the sample will be very cumbersome. So we can divide n life data into k groups according to a certain time interval (or the number of operations) and take the medium value of failure time ranges of each group as the approximate value of each life data in this group. Now the life variance s^2 and the standard deviation of lives s of the sample can be calculated through the formula (2-28):

$$s^2 = \frac{1}{n-1} \sum_{i=1}^{k} \Delta m_i (t_{zi} - \bar{t})^2 \qquad (2\text{-}31)$$

$$s = \sqrt{\frac{1}{n-1} \sum_{i=1}^{k} \Delta m_i (t_{zi} - \bar{t})^2} \qquad (2\text{-}32)$$

where, Δm_i = frequency of group i

t_z = the medium value of failure time range in group i

When the sample is becoming larger, (i.e. n is gradually increasing), though the life variance s^2 and the standard deviation of lives s of the sample fluctuates, yet the general trend is respectively tending to be a stable value. The stable value is the population of the life variance and the standard deviation of lives. In practice, the population of the life variance and the standard deviation of lives can be estimated through the life variance and the standard deviation of lives of the sample.

3.1.6 RELIABLE LIFE

The reliability function $R(t)$ is the function of the operating time t of the product. The operating time needed to decrease the reliability of the product to a specified value R is called the reliable life of the product. It's represented with the sign t_R. The reliable life can be represented by the following mathematical relation, i.e.

$$R(t_R) = R \qquad (2\text{-}33)$$

The relationship between the reliable life t_R and the reliability R can be represented by Figure 2-5.

3.1.7 MEDIAN LIFE

When the reliability of the product is equal to 0.5, the reliable life is called the median life. It's represented with the sign $t_{0.5}$. The physical meaning of the median life is the operating time needed when half of the products failure is detected in a batch of products. The median life can be represented by the formula (2-34), i.e.

$$R(t_{0.5}) = 0.5 \qquad (2\text{-}34)$$

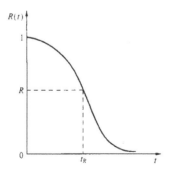

Figure 2-5 The relationship between reliable life t_R and the reliability R

3.1.8 SUCCESS RATE

Success rate refers to the probability that the product completes specified functions under stated conditions or the probability of the success that the product tests under certain conditions.

3.2 RELIABILITY CHARACTERISTIC PARAMETERS OF REPAIRABLE PRODUCTS

The characteristic parameters of repairable products mainly include: mean fault rate, mean time between fault, validity, mean repair time, maintenance costs rate, reliability and fault density.

3.2.1 MEAN FAULT RATE λ

It evaluates the fault-free feature or the frequency of fault of the equipment. It refers to the ratio between the number of product fault and the total number of products under stated conditions for a specified period of time. The calculation formula of the observed value is:

$$\lambda = \frac{v}{nt} \qquad (2\text{-}35)$$

where, n is the number of test samples, t is the test time while v is the number of fault during the test.

In order to distinguish various types of fault, we uniformly converted to various types of fault (fatal fault, serious fault and minor fault) as the general fault. The fault rate D represents the frequency of fault.

$$D = K \frac{\sum_{i=1}^{4} \varepsilon_i v_i}{nt} \qquad (2\text{-}36)$$

where, v_i is the number of class i fault of the test sample occurred, ε_i is the hazard coefficient of class i fault ($\varepsilon_1 = 100$ is recommended to represent fatal fault, ε_2 =5 represents the serious fault, ε_3 =1 represents the general fault, ε_4 =0.2 represents the minor fault.); n is the number of test samples. t is the test time. K is the time coefficient of the reliability test.

3.2.2 MEAN TIME BETWEEN FAILURE (*MTBF*)

The expression of the mean time between failure (*MTBF*), that is, the mean duration during which the product can maintain specified functions under stated conditions, the formula is

$$MTBF = \frac{1}{N} \sum_{i=1}^{N} Ti \qquad (2\text{-}37)$$

where, N is the total number of equipments undertaking the test; Ti is the time between failure of the equipment i.

3.2.3 VALIDITY *A*

The validity A is the overall evaluation for low voltage switchgear. It refers to the probability to maintain its required functions under certain stated conditions for a specified period of time.

$$A = \frac{\sum_{i=1}^{N} tw_i}{\sum_{i=1}^{N} (tw_i + tr_i)} \qquad (2\text{-}38)$$

where, N refers to the total number of switchgear undertaking the test. tw_i is the operating time of the switchgear i during the period of test. tr_i refers to the maintenance time (including diagnosis, preparation, servicing and trial run period) of switchgear i during the period of test.

3.2.4 MEAN TIME TO REPAIR (*MTTR*)

Mean time to repair (*MTTR*) is mainly used for assessing and evaluating the servicing level of the stand-alone system.

$$MTTR = \frac{\sum_{i=1}^{N} t_i}{N} \qquad (2\text{-}39)$$

where, N refers to the total number of stand-alone systems repaired due to fault. t_i refers to the time for repairing the stand alone system i.

3.2.5 REPAIR RATE

Repair rate refers to the probability to complete repair in unit time after the repair time of the product reaches t. It's represented by $m(t)$.

3.2.6 RELIABILITY $R(T)$

Reliability at time t refers to the fault free probability in the time interval $(0, t)$ when the equipment is launched normally.

3.2.7 FAILURE DENSITY $F(T)$

The failure density $f(t)$ refers to the probability that the equipment malfunction for the first time during the period of $(t, t+\Delta t)$.

3.3 THE RELATIONSHIP BETWEEN THE FAILURE DENSITY FUNCTION AND THE CHARACTERISTIC QUANTITY OF RELIABILITY

(1) The relationship between the cumulative failure probability and the cumulative failure distribution function

The cumulative failure distribution function and the cumulative failure probability refer to the same function, thus both represented by $F(t)$.

(2) The relationship between the cumulative failure probability $F(t)$ and the reliability function $R(t)$.

$$F(t) = 1 - R(t) \qquad (2\text{-}40)$$

(3) The relationship between the cumulative failure distribution function $F(t)$, the reliability function $R(t)$ and the failure density function $f(t)$

$$F(t) = \int_0^t f(t)dt \qquad (2\text{-}41)$$

or $f(t) = \dfrac{dF(t)}{dt} = F'(t)$ (2-42)

$R(t) = 1 - F(t) = \displaystyle\int_{t}^{\infty} f(t)dt$ (2-43)

or $f(t) = \dfrac{d[1 - R(t)]}{dt} = -\dfrac{dR(t)}{dt} = -R'(t)$ (2-44)

The relationship between the cumulative failure distribution function $F(t)$, the reliability function $R(t)$ and the failure density function $f(t)$ can be represented by Figure 2-6.

(4) The relationship between the reliability function $R(t)$ and the failure rate function $\lambda(t)$

$\lambda(t) = \dfrac{f(t)}{R(t)}$ (2-45)

or $\lambda(t) = -\dfrac{R'(t)}{R(t)}$ (2-46)

or $R(t) = e^{-\int_{0}^{t} \lambda(t)dt}$ (2-47)

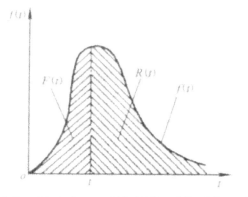

Figure 2-6   The relationship between $F(t)$, $R(t)$ and the failure density function $f(t)$

(5) The relationship between the cumulative failure distribution function $F(t)$, the failure density function $f(t)$ and the failure rate function $\lambda(t)$

$F(t) = 1 - e^{-\int_{0}^{t} \lambda(t)dt}$ (2-48)

$$f(t) = \lambda(t)e^{-\int_0^t \lambda(t)dt}$$
(2-49)

(6) The relationship between the average life μ of the population and the failure density function $f(t)$

$$\mu = \int_0^\infty tf(t)dt$$
(2-50)

(7) The relationship between the standard deviation of lives of the population σ *and the* failure density function $f(t)$,

The relationship between the life variance of the population and the failure density function $f(t)$ is

$$\sigma^2 = \int_0^\infty (t-\mu)^2 f(t)dt$$
(2-51)

or $\sigma^2 = \int_0^\infty t^2 f(t)dt - \mu^2$
(2-52)

The relationship between the standard deviation of lives σ of the population and the failure density function $\lambda(t)$ refers to

$$\sigma = \sqrt{\int_0^\infty (t-\mu)^2 f(t)dt}$$
(2-53)

or $\sigma = \sqrt{\int_0^\infty t^2 f(t)dt - \mu^2}$
(2-54)

We can make a table to summarize the mathematical expressions of all the above reliability characteristic parameters and the relationships between characteristic parameters, and between them and the failure density function $f(t)$, as shown in Table 2-1.

4   FAILURE DISTRIBUTION TYPES

4.1 COMMON FAILURE DISTRIBUTION TYPES

The distribution type of one random variable refers to the function type of the density function or cumulative distribution function of the random variable. Exponential distribution is the most common distribution type in the reliability theory. Besides, other common distribution types include Weibull distribution, normal distribution and lognormal distribution and so on.

Table 2-1 Mathematical expressions and correlations of reliability characteristic parameters

Names and signs of reliability characteristic parameters	Mathematical expressions	Correlations
Cumulative failure probability $F(t)$	$F(t) = \begin{cases} P(L < t) & (t \geq 0) \\ 0 & (t < 0) \end{cases}$	$F(t) = \int_0^t f(t)dt$ $f(t) = \dfrac{dF(t)}{dt} = F'(t)$
Reliability function $R(t)$	$R(t) = \begin{cases} P(L \geq t) & (t \geq 0) \\ 1 & (t < 0) \end{cases}$	$R(t) = 1 - F(t) = \int_t^\infty f(t)dt$ $R(t) = 1 - F(t)$ $f(t) = -\dfrac{dR(t)}{dt} = -R'(t)$
Failure rate function $\lambda(t)$	$\lambda(t) = \dfrac{m(t + \Delta t) - m(t)}{[n - m(t)]\Delta t}$	$\lambda(t) = \dfrac{f(t)}{R(t)}$ $\lambda(t) = -\dfrac{R'(t)}{R(t)}$ $R(t) = e^{-\int_0^t \lambda(t)dt}$ $F(t) = 1 - e^{-\int_0^t \lambda(t)dt}$ $f(t) = \lambda(t)e^{-\int_0^t \lambda(t)dt}$
Average life of the sample $\bar{t}$ Average life of the population	$\bar{t} = \dfrac{1}{n}\sum_{i=1}^{n} t_i$ $\bar{t} = \dfrac{1}{n}\sum_{i=1}^{k} t_{zi}\Delta m_i$	$\mu = \int_0^\infty tf(t)dt$
Standard deviation of lives of the sample s Standard deviation of lives of the population σ	$s = \sqrt{\dfrac{1}{n-1}\sum_{i=1}^{n}(t_i - \bar{t})^2}$ $s = \sqrt{\dfrac{1}{n-1}\sum_{i=1}^{k}\Delta m_i(t_{zi} - \bar{t})^2}$	$\sigma = \sqrt{\int_0^\infty (t - \mu)^2 f(t)dt}$ $\sigma = \sqrt{\int_0^\infty t^2 f(t)dt - \mu^2}$
Reliable life t_R	$R(t_R) = R$	
Median life $t_{0.5}$	$R(t_{0.5}) = 0.5$	

4.1.1 EXPONENTIAL DISTRIBUTION

Exponential distribution (especially the single-parameter exponential distribution) plays a very important role in reliability technology. The service lives of many products follow exponential distribution; and it is quite easy to calculate the characteristic parameters such as the average life and the failure rate via simple and accurate formulas under exponential distribution. So, in current standards made by many countries, most of them are identifying the reliability level of electronic components products on the basis of exponential distribution. The sampling table for qualification tests in Chinese national standards GB1772 *the Test Method for Failure Rate of Electronic Components* and GB/T 15510-2008 *General Rules for Reliability test of Electromagnetic Relay for Control Circuits* are achieved on the condition that the service life follows the exponential distribution.

Exponential distribution can be divided into single-parameter exponential distribution and two-parameter exponential distribution. Herein, the single-parameter exponential distribution refers to the exponential distribution mentioned in current materials. It just has one parameter, which is a special situation among two-parameter exponential distribution. Thus, the two-parameters exponential distribution can also be called exponential distribution of common forms.

4.1.1.1   SINGLE-PARAMETER EXPONENTIAL DISTRIBUTION

The density function of L (it refers to the random variable "product life"), and the failure density function is

$$f(t) = \begin{cases} \lambda e^{-\lambda t} & (t \geq 0) \\ 0 & (t < 0) \end{cases} \tag{2-55}$$

Then, the random variable L is considered to follow the single-parameter exponential distribution. Herein, the parameter λ refers to the failure rate under the single-parameter exponential distribution.

Its cumulative failure distribution function is

$$F(t) = \begin{cases} 1 - e^{-\lambda t} & (t \geq 0) \\ 0 & (t < 0) \end{cases} \tag{2-56}$$

Figures of the failure density function $f(t)$ of the single-parameter exponential distribution and the cumulative failure distribution function $F(t)$ are shown as Figure 2-7 and Figure 2-8.

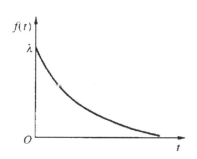

Figure 2-7

Failure density function of the single-parameter exponential distribution

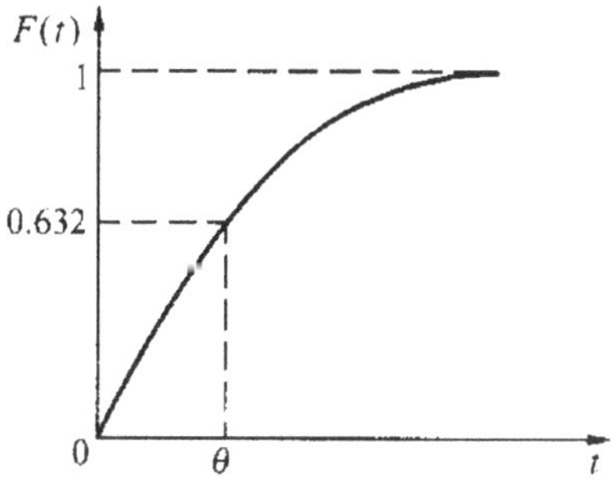

Figure 2-8

Cumulative failure distribution function of the single-parameter exponential distribution

Expression of reliability characteristic parameters of the single-parameter exponential distribution is as follows;

(1) The reliability function $R(t)$ is

$$R(t) = 1 - F(t) = \begin{cases} e^{-\lambda t} & (t \geq 0) \\ 1 & (t < 0) \end{cases} \tag{2-57}$$

The reliability function of single-parameter exponential distribution is shown in Figure 2-9.

(2) The failure rate function $\lambda(t)$ is

$$\lambda(t) = \frac{f(t)}{R(t)} = \begin{cases} \lambda & (t \geq 0) \\ 0 & (t < 0) \end{cases} \tag{2-58}$$

(3) Average life θ (average life is usually represented by θ when the exponential distributions) is

$$\theta = \int_0^\infty t f(t) dt = \frac{1}{\lambda} \tag{2-59}$$

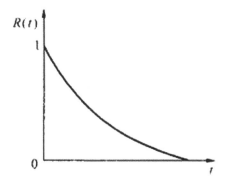

Figure 2-9

The reliability function of single-parameter exponential distribution

(4) Life variance σ^2 and standard deviation of lives σ refer to

$$\sigma^2 = \int_0^\infty t^2 f(t)dt - \mu^2 = \theta^2 \qquad (2\text{-}60)$$

$$\sigma = \frac{1}{\lambda} = \theta \qquad (2\text{-}61)$$

(5) Reliable life t_R refers to

$$t_R = -\frac{1}{\lambda}\ln R = \frac{1}{\lambda}\ln\frac{1}{R} \qquad (2\text{-}62)$$

(6) Median life $t_{0.5}$ refers to

$$t_{0.5} = -\frac{1}{\lambda}\ln 0.5 = \frac{\ln 2}{\lambda} = \frac{0.693}{\lambda} = 0.693\theta$$

$$(2\text{-}63)$$

Characteristics of exponential distributions with single parameter:

(1) If the product life follows the single-parameter exponential distribution, then the failure rate function is equal to a constant λ; on the contrary, if we know $t \geq 0$, the failure rate of the product is a constant λ, then the product life is bound to follow the single-parameter exponential distribution.

(2) When the product life follows the single-parameter exponential distribution, its average life θ and the failure rate λ are reciprocal with each other.

(3) When the product life follows the single-parameter exponential distribution, its standard deviation of life σ is equal to average lifeθ , and is inversely proportional to the failure rateλ.

4.1.1.2   TWO-PARAMETERS EXPONENTIAL DISTRIBUTION

If the density functions of random variable L is

$$f(t) = \begin{cases} \lambda e^{-\lambda(t-v)} & (t \geq v \geq 0) \\ 0 & (t < v) \end{cases} \qquad (2\text{-}64)$$

Then, the random variable L is considered to follow the two-parameters exponential distribution.

Its cumulative failure distribution function is

$$F(t) = \begin{cases} 1 - e^{-\lambda(t-v)} & (t \geq v \geq 0) \\ 0 & (t < v) \end{cases} \qquad (2\text{-}65)$$

where λ refers to the failure rate v refers to the position parameter, which indicates that the product won't occur failure when $t < v$.

Single-parameter exponential distribution refers to the special case of two-parameters exponential distribution when $v = 0$. Thus, two-parameter exponential distribution can also be called exponential distributions of common forms. Figures of the failure density function and cumulative failure distribution function of two-parameters exponential distribution are shown as Figure 2-10 and Figure 2-11.

Expressions of reliability characteristic parameters of two-parameters exponential distribution are as follows:

(1)   The expression of reliability function R (t) is

$$R(t) = 1 - F(t) = \begin{cases} e^{-\lambda(t-v)} & (t \geq v) \\ 1 & (t < v) \end{cases} \qquad (2\text{-}66)$$

The curve of reliability function of two-parameters exponential distribution is as shown in Figure 2-12.

(2)   Failure rate function λ (t)

$$\lambda(t) = \frac{f(t)}{R(t)} = \begin{cases} \lambda & (t \geq v) \\ 0 & (t < v) \end{cases} \qquad (2\text{-}67)$$

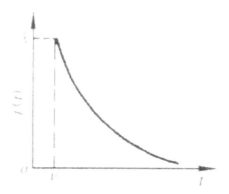

Figure 2-10 The failure density function of two- parameter exponential distribution

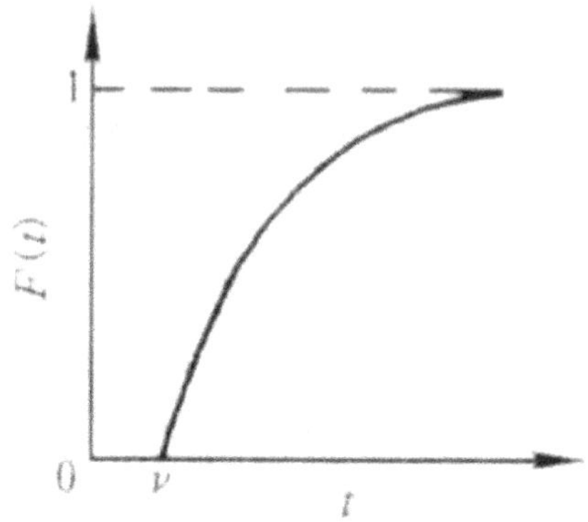

Figure 2-11 The cumulative failure distribution function of two-parameter exponential distribution

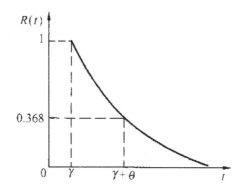

Figure 2-12 The reliability function of two-parameters exponential distribution

(3) Average life θ

$$\theta = \int_0^\infty t\lambda e^{-\lambda(t-v)}dt = \frac{1}{\lambda} + v \qquad (2\text{-}68)$$

(4) Life variance σ^2 and life standard deviation σ

$$\sigma^2 = \int_0^\infty t^2 \lambda e^{-\lambda(t-v)}dt - \theta^2 = \frac{1}{\lambda^2} = (\theta - v)^2 \qquad (2\text{-}69)$$

$$\sigma = \frac{1}{\lambda} = \theta - \nu \qquad (2\text{-}70)$$

(5) Reliable life t_R

$$t_R = \nu - \frac{\ln R}{\lambda} = \nu + \frac{1}{\lambda}\ln\frac{1}{R} \qquad (2\text{-}71)$$

(6) Median life $t_{0.5}$

$$t_{0.5} = \nu - \frac{\ln 0.5}{\lambda} = \nu + \frac{\ln 2}{\lambda} = 0.307\nu + 0.693\theta \qquad (2\text{-}72)\cdot$$

Characteristics of two-parameter exponential distribution:

(1) If the product life follows the two-parameter exponential distribution, then the failure rate function is equal to the constant λ when $t \geq \nu$, and yet equal to zero when $t < \nu$. On the contrary, if $t \geq \nu$, the failure rate of the product is equal to the constant λ, and failure rate is equal to zero when $t < \nu$, then the product life is bound to follow the two-parameter exponential distribution.

(2) If the product life follows the two-parameter exponential distribution, then the average life θ and the failure rate λ are not reciprocal with each other anymore. It should be $\theta = \frac{1}{\lambda} + \nu$.

(3) If the product life follows the two-parameter exponential distribution, then its life standard deviation σ and the failure rate λ are still reciprocal with each other, but the σ is no longer equal to the average life. It should not $\sigma = \theta - \nu$.

4.1.2 WEIBULL DISTRIBUTION

Weibull distribution is the most complicated distribution usually used in the reliability theory. Weibull distribution includes three parameters, so it fits greatly for various types of test data. Thus it has been widely used in reliability technology.

4.1.2.1 DEFINITION OF WEIBULL DISTRIBUTION

If the density function of random variable is L, i.e. the failure density function is

$$f(t) = \begin{cases} \dfrac{m}{t_0}(t-v)^{m-1} e^{-\frac{(t-v)^m}{t_0}} & (t \geq v) \\ 0 & (t < v) \end{cases} \qquad (2\text{-}73)$$

Then the random variable L is considered to follow Weibull distribution. Its cumulative failure distribution function is

$$F(t) = \begin{cases} 1 - e^{-\frac{(t-v)^m}{t_0}} & (t \geq v) \\ 0 & (t < v) \end{cases} \qquad (2\text{-}74)$$

where m refers to shape parameter; t_0 refers to scale parameter; v refers to position parameter.

4.1.2.2   THE MEANING OF THREE PARAMETERS OF WEIBULL DISTRIBUTION

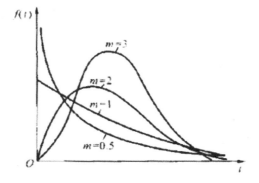

Figure 2-13 Failure density curves with different m values when $v = 0$、$t_0 = 1$

(1) Shape parameter m. The curve shapes of the failure density, reliability, cumulative failure distribution and failure rate of Weibull distribution vary with the m value. Thus m is called the shape parameter. Wherein, the failure density curve is influenced by the m value most significantly. When the values of position parameter v and scale parameter t_0 are fixed (i.e. $v = 0$、$t_0 = 1$), failure density curves with different m values are shown in Figure 2-13.

From Figure 2-13 we can see that failure density curves with different m values can be divided into three types:

1) When $m < 1$, the curve $f(t)$ decreases monotonically with the time, and the curve $f(t)$ will never intersect with the ordinate axis (take the ordinate axis as its asymptotic line).

2) When $m = 1$, the curve $f(t)$ is exponential curve, it decreases monotonically with the time and intersects with the ordinate axis.

3) When $m > 1$, the curve $f(t)$ will reach its peak value with the increase of time, then decreases and tends to be zero. The curve $f(t)$ is shown as single-peak types.

(2) Position parameter v. When m-value and t_0 value are fixed (for example, let $t_0 = 1$、$m = 2$), the failure density curve with different v values is shown as Figure 2-14.

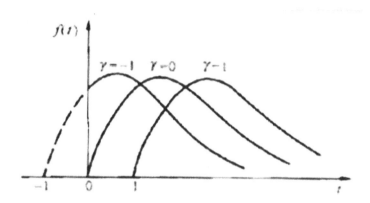

Figure 2-14 Failure density curves with different m values when $t_0 = 1$、$m = 2$

From the Figure 2-14 we can see shapes of failure density curves with different v values are identical, while only their positions in coordinate system vary.

When $v < 0$, the position of the curve $f(t)$ will move to the left in parallel for the distance $|v|$ at the position $v = 0$; When $v > 0$, the position of the curve $f(t)$ will move to the right in parallel for the distance $|v|$ at the position $v = 0$. So v is called the position parameter.

(3) Scale parameter t_0 When m value and v value are fixed, heights and widths of failure density curves of Weibull distribution with different t_0 values are both different. Failure density curves is shown as Figure 2-15 with different t_0 values

when $m = 2, v = 0$. From Figure 2-15 we can see that the height of failure density curve is smaller while the width is larger as the t_0-value increases.

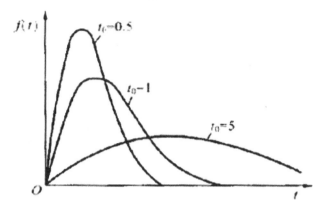

Figure 2-15   Failure density curves with different t_0-values when $m = 2, v = 0$

True scale parameter η

$$\eta = t_0^{\frac{1}{m}} \qquad\qquad (2\text{-}75)$$

When $v = 0$, we shall get

$$\eta f(t) = \begin{cases} mt'^{m-1}e^{-t'^m} & (t' \geq 0) \\ 0 & (t' < 0) \end{cases} \qquad\qquad (2\text{-}76)$$

where, $t' = t/\eta$

i.e. when the m and v values are same, but different η value, their failure density curves will be completely coincided only if the scales of the ordinate and the abscissa are changed properly (enlarge the ordinate scale by η times and reduce the abscissa scale by η times).Thus, η is called the true scale parameter (also be called characteristic life).

4.1.2.3   CHARACTERISTIC PARAMETERS UNDER WEIBULL DISTRIBUTION

(1) Reliability function $R(t)$

$$R(t) = 1 - F(t) = \begin{cases} e^{-\frac{(t-v)^m}{t_0}} & (t \geq v) \\ 1 & (t < v) \end{cases} \qquad (2\text{-}77)$$

Reliability curves with different m values when $t_0 = 1$ and $v = 1$ are shown in Figure 2-16.

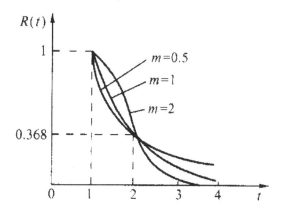

Figure 2-16 Reliability curves with different *m*-values when $t_0 = 1$ and $v = 1$

(2) Failure rate function $\lambda(t)$

When $t \geq v$, $\lambda(t) = \dfrac{f(t)}{R(t)} = \dfrac{\dfrac{m}{t_0}(t-v)^{m-1}e^{-\frac{(t-v)^m}{t_0}}}{e^{-\frac{(t-v)^m}{t_0}}} = \dfrac{m}{t_0}(t-v)^{m-1}$

Thus, the failure density function of Weibull distribution is

$$\lambda(t) = \begin{cases} \dfrac{m}{t_0}(t-v)^{m-1} & (t \geq v) \\ 0 & (t < v) \end{cases} \qquad (2\text{-}78)$$

Failure rate curves with different m values are shown in Figure 2-17.

(3) Average life

$$\mu = \eta\Gamma\left(1 + \frac{1}{m}\right) + v \qquad (2\text{-}79)$$

where $\Gamma\left(1+\dfrac{1}{m}\right)$: it can be found in the Gamma function table(Appendix 1) of

Mathematics manuals according to m values($\displaystyle\int_0^\infty x^{p-1}e^{-x}dx(p>0)$ is

referred to as the Gamma function, and represented by symbol Γ (P)). However, in Gamma function table, only Γ (P) values under P = 1 ~ 2 are listed. When

$1+\dfrac{1}{m}$ exceeds 2, $\Gamma\left(1+\dfrac{1}{m}\right)$ can be calculated through the following natures

of Gamma functions, i.e.

$$\Gamma(P+1)=P\Gamma(P)\quad(P>0)\tag{2-80}$$

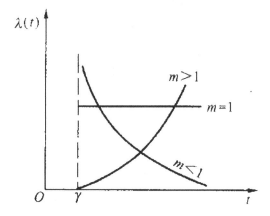

Figure 2-17 Failure rate curves with different *m* values

(4) Standard deviation of life σ

$$\sigma^2=\eta^2\left[\Gamma\left(1+\frac{2}{m}\right)-\Gamma^2\left(1+\frac{1}{m}\right)\right]\tag{2-81}$$

$$\sigma=\eta\left[\Gamma\left(1+\frac{2}{m}\right)-\Gamma^2\left(1+\frac{1}{m}\right)\right]^{\frac{1}{2}}\tag{2-82}$$

where, $\Gamma^2\left(1+\dfrac{1}{m}\right)$: represents $\left[\Gamma\left(1+\dfrac{1}{m}\right)\right]^2$;

$\Gamma\left(1+\dfrac{2}{m}\right)$: can be calculated through Gamma function table according to m value.

(5) Reliable life t_R

$$t_R = v + \eta(-\ln R)^{\frac{1}{m}} \qquad (2\text{-}83)$$

(6) Median life $t_{0.5}$

$$t_{0.5} = v + \eta(-\ln 0.5)^{\frac{1}{m}} = v + \eta(\ln 2)^{\frac{1}{m}} = v + \eta(0.693)^{\frac{1}{m}} \qquad (2\text{-}84)$$

4.1.2.4   CHARACTERISTICS OF WEIBULL DISTRIBUTION

(1) Weibull distribution can be divided into two types: the Weibull distribution is called two-parameter Weibull distribution when $v = 0$; the Weibull distribution is called three-parameter Weibull distribution when $v \neq 0$.

(2) When the shape parameter $m = 1$, the failure density function of three-parameter Weibull distribution is

$$f(t) = \begin{cases} \dfrac{1}{t_0} e^{-\frac{t-v}{t_0}} & (t \geq v) \\ 0 & (t < v) \end{cases} \qquad (2\text{-}85)$$

Two-parameter Weibull distribution is a special situation of three-parameter Weibull distribution.

(3) When the shape parameter $m = 3\sim4$, the failure density curve of Weibull distribution will be coincided with failure density curve of normal distribution as shown in Figure 2-18. In this Figure, the dash line refers to the failure density curve under normal distribution (its average life $\mu = 0.8963$, standard deviation of life $\sigma = 0.303$); the solid line refers to the failure density curve of Weibull distribution (its parameters are as follows: $m = 3.25$, $\eta = 1$, $v = 0$)

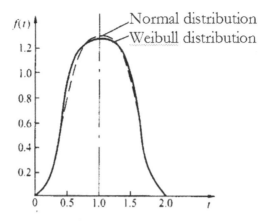

Figure 2-18 Comparison between Weibull distribution and normal distribution

4.1.3 NORMAL DISTRIBUTION

Normal distribution is used in reliability technology frequently. It's mostly used in describing the failure of products caused by attrition and degradation.

4.1.3.1 DEFINITION OF NORMAL DISTRIBUTION

If the density function of the random variable X is

$$f(x) = \frac{1}{\sqrt{2\pi}\sigma} e^{-\frac{(x-\mu)^2}{2\sigma^2}} \qquad (-\infty < x < \infty) \qquad (2\text{-}86)$$

where μ refers to the position parameter of normal distribution; σ refers to the scale parameter of normal distribution. Then the random variable X follows the normal distribution with parameters of μ and σ, referred as $X \sim N (\mu, \sigma^2)$. The cumulative distribution function of the normal distribution is

$$F(x) = \frac{1}{\sqrt{2\pi}\sigma} \int_{-\infty}^{x} e^{-\frac{(v-\mu)^2}{2\sigma^2}} dv \qquad (-\infty < x < \infty) \qquad (2\text{-}87)$$

The density curve of normal distribution is shown in Figure 2-19.

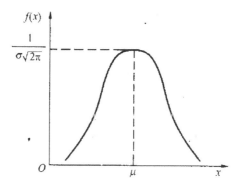

Figure 2-19 The density curve of normal distribution

4.1.3.2   MEANINGS OF THE TWO PARAMETERS μ AND σ IN NORMAL DISTRIBUTION

(1) Position parameter μ

When the random variable X follows normal distribution, the mean value $E(x)$ of X is

$$E(X) = \int_{-\infty}^{\infty} x \frac{1}{\sqrt{2\pi}\sigma} e^{-\frac{(x-\mu)^2}{2\sigma^2}} dx = \mu \tag{2-88}$$

Thus, the position parameter μ in normal distribution refers to the mean value of the random variable X. When the random variable refers to the life, the position parameter μ is equal to the mean life; it represents the central position of normal distribution.

(2) Scale parameter σ

When the random variable X follows normal distribution, the variance $D(x)$ of X is

$$D(X) = \int_{-\infty}^{\infty} (x - \mu)^2 \frac{1}{\sqrt{2\pi}\sigma} e^{-\frac{(x-\mu)^2}{2\sigma^2}} dx = \sigma^2 \tag{2-89}$$

$$\sqrt{D(X)} = \sigma \tag{2-90}$$

Thus, the scale parameter σ in normal distribution refers to the standard deviation of the random variable X. When the random variable refers to the life, the scale parameter σ is equal to the standard deviation of life.

Density function curves of three different normal distributions with the same position parameter μ and different scale parameters σ are shown in Figure 2-20.

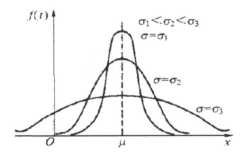

Figure 2-20 Density function curves of three different normal distributions with the same position parameter μ and different scale parameters σ

4.1.3.3   CHARACTERISTICS OF NORMAL DISTRIBUTION

(1) The density curve of normal distribution has a bell-like shape, and is symmetric to the line $x = \mu$.

(2) The position parameter μ is the mean value of the random variable X. It refers to the central position of the density curve of normal distribution. The scale parameter refers to the standard deviation of the random variable X; it reflects the discrete degree of normal distribution.

(3) The density function $f(x)$ achieves its maximum value $\dfrac{1}{\sqrt{2\pi}\sigma}$ when $x = \mu$; thus the value of the probability is larger when the random variable X approaches μ. On the contrary, when the value of the random variable X is further away from μ, the probability is smaller.

(4) When $\mu = 0$, $\sigma = 1$, the normal distribution is called standard normal distribution. Its density function and cumulative distribution function used to be represented by $\varphi(z)$ and $\Phi(z)$, i.e.

$$\varphi(z) = \frac{1}{\sqrt{2\pi}} e^{-\frac{z^2}{2}} \quad (-\infty < z < \infty) \tag{2-91}$$

$$\Phi(z) = \int_{-\infty}^{z} \frac{1}{\sqrt{2\pi}} e^{-\frac{v^2}{2}} dv \, (-\infty < z < \infty) \tag{2-92}$$

where, v: the symbol of variable

The standard normal distribution is usually referred to as N (0,1). Its density curve is shown in Figure 2-21.

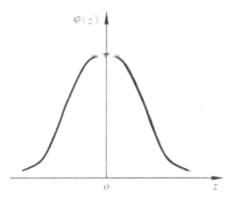

Figure 2-21   The density curve of standard normal distribution

Table 2-2 Summary of Failure density curves and reliability curves of the several failure distribution types used commonly

Distribution Type	Failure Density Curves	Reliability Curves
Single-Parameter Exponential Distribution		
Two-Parameters Exponential Distribution		
Weibull Distribution		
Normal Distribution		

Table 2-3 Group statistics of life data

Group number i	Life interval	Failure number Δm_i	Frequency f_i^*
1	$a_0 \sim a_1$	Δm_1	f_1^*
2	$a_1 \sim a_2$	Δm_2	f_2^*
$\vdots$	$\vdots$	$\vdots$	$\vdots$
k	$a_{k-1} \sim a_k$	Δm_k	f_k^*

4.2 ESTIMATION METHOD OF FAILURE DISTRIBUTION TYPES

If there are previous empirical data for the product, then the failure distribution type can be assumed to base on it. If there are no such empirical data, we can select certain amount of samples and perform the test, and then draw the failure frequency histogram or the reliability function figure according to test data. Compare these Figures with failure density function Figures and reliability function Figures of various commonly used failure distribution types in Table 2-2, then estimate the failure distribution type.

4.2.1 ESTIMATION METHOD OF FAILURE DISTRIBUTION TYPES FOR LARGE SAMPLE

The large sample means the number of life test data is relatively huge. Now the life data should be divided according to certain intervals (or the number of operations), and the process is as follows:

(1) Dividing the life data into k groups.

(2) Calculating the failure number Δm_i and frequency $f_i^* \left(f_i^* = \dfrac{\Delta m_i}{n} \right)$ of each

group and make a table as shown in Table 2-3.

(3) To draw the failure frequency histogram according to the data in Table 2-3.

(4) Roughly draw the failure density curve according to the failure frequency histogram.

(5) To compare the shape of the failure density curve with shapes of the failure density curves of various distribution types, to estimate the failure distribution type of product.

4.2.2 ESTIMATION METHOD OF FAILURE DISTRIBUTION TYPES FOR SMALL SAMPLE

The number of life test data for small samples is very little. Draw the reliability curve to estimate the failure distribution type. Its detailed processes are as follows:

(1) Calculating the reliability function $R(t_i)$ when $t = t_i$, when the number of test samples $n > 20$

$$R(t_i) = 1 - \frac{i}{n} \qquad (2\text{-}93)$$

When $n \leq 20$ $R(t_i) = 1 - \frac{i - 0.5}{n}$ (2-94)

or $R(t_i) = 1 - \frac{i}{n+1}$ (2-95)

or $R(t_i) = 1 - \frac{i - 0.3}{n + 0.4}$ (2-96)

where t_i refers to the life data of failure sample i ($i = 1,2,\ldots, r$).

(2) Making a table for i, t_i, $R(t_i)$, as shown in Table 2-4.

Table 2-4 Reliability function values

i	1	2	...	r
t_i	t_1	t_2	...	t_r
$R(t_i)$	$R(t_1)$	$R(t_2)$	...	$R(t_r)$

(3) Plotting points according to $[t_i, R(t_i)]$ in the rectangular coordinate and draw the reliability function figure .

(4) Comparing the reliability function figure with the shapes of the reliability curves of various failure distribution types in Table 2-2, to estimate the failure distribution type of product.

4.3 VERIFICATION METHOD OF FAILURE DISTRIBUTION TYPES

4.3.1 GRAPHICAL VERIFICATION METHOD OF FAILURE DISTRIBUTION TYPES

When we have estimated the failure distribution type of the product through the above method, we can verify the correctness of the estimation by graphical verification method. If the distribution type is supposed to be exponential distribution, then a piece of unilateral logarithm graph paper should be used for verification. If the distribution type is supposed to be Weibull distribution, then a piece of Weibull probability paper should be used for verification. If the failure distribution type is supposed to be normal distribution, then a piece of normal probability paper should be used for verification.

4.3.1.1 UTILIZE A PIECE OF UNILATERAL LOGARITHM GRAPH PAPER TO VERIFY WHETHER THE FAILURE DISTRIBUTION TYPE IS EXPONENTIAL DISTRIBUTION

(1) Unilateral logarithm graph paper

When the life follows the single-parameter exponential distribution, take $\ln R(t)$ as the ordinate and t as the abscissa, a rectangular coordinate system with uniform scales is formed. If add the $R(t)$ coordinate on the ordinate $\ln R(t)$ according to the relationship of $R(t) = e^{\ln R(t)}$, then a piece of unilateral logarithm graph paper is formed. The $R(t)$ coordinate is logarithmic coordinate with non-uniform scales, while scales of the $\ln R(t)$ coordinate and t coordinate are both uniform. The ordinate of the practical unilateral logarithm graph paper has no uniform scales but with logarithmic relationship scales, while the abscissa has uniform scales. It's as shown in Figure 2-22.

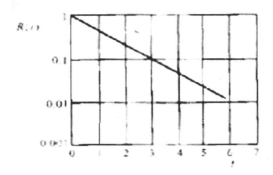

Figure2-22 Unilateral logarithm graph paper

(2) Graphical verification method

According to life test data, points plotted in coordinate system t-$R(t)$ on a piece of unilateral logarithm graph paper are approximated in accordance with $[t_i, R(t_i)]$ on a straight line which passes the point $[t = 0, R(t) = 1]$, then the failure distribution type can be judged to be single-parameter exponential distribution.

According to life test data, points plotted in coordinate system t-$R(t)$ on a piece of unilateral logarithm graph paper are approximated in accordance with $[t_i, R(t_i)]$ on a straight line which doesn't pass the point $[t = 0, R(t) = 1]$, then the failure distribution type can be judged to be two-parameters exponential distribution.

4.3.1.2 UTILIZE A PIECE OF WEIBULL PROBABILITY PAPER TO VERIFY WHETHER THE FAILURE DISTRIBUTION TYPE IS EXPONENTIAL DISTRIBUTION

(1) Structural principles of Weibull probability paper

Let
$$\left.\begin{aligned} &ln\,ln\frac{1}{1 - F(t)} = Y \\ &ln\,t = X \\ &ln\,t_0 = B \end{aligned}\right\} \tag{2-97}$$

Then
$$\left.\begin{aligned} &t = e^X \\ &F(t) = 1 - e^{-e^Y} \\ &t_0 = e^B \end{aligned}\right\} \tag{2-98}$$

Take X as the abscissa and Y as the ordinate in common rectangular coordinates system, and add the t coordinate on the abscissa and add the $F(t)$ coordinate on the ordinate according to the relation (2-98), a schematic diagram of Weibull probability paper is formed as shown in Figure 2-23.

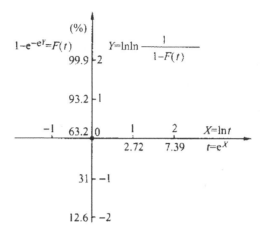

Figure 2-23   Schematic diagram of Weibull probability paper

In Figure 2-23, scales of the $F(t)$ coordinate and the t coordinate are both non uniform. In order to facilitate practical use, four points on the ordinate and the abscissa ($F(t)$ coordinate, Y coordinate, X coordinate, t coordinate) in Figure 2-23 are usually marked on four sides, i.e. the left, the right, the top and the bottom of the Weibull probability paper as shown in Figure 2-24. The point of $X = 1$, $Y = 0$ in the Weibull probability paper is called the estimation point of m.

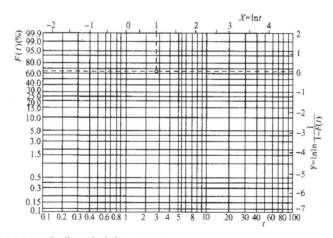

Figure 2-24   Weibull probability paper

(2) The graphical verification method of the Weibull distribution when $v = 0$ is as follows:

1) Using the following formulas to calculate the cumulative failure probability $F(t_i)$ on the basis of life test data $(t_1, t_2, ..., t_n)$ and the total number of test sample n

When $n > 20$

$$F(t_i) = \frac{i}{n} \tag{2-99}$$

When $n \leq 20$,

$$F(t_i) = \frac{i - 0.5}{n} \tag{2-100}$$

$$\text{or } F(t_i) = \frac{i}{n + 1} \tag{2-101}$$

$$\text{or } F(t_i) = \frac{i - 0.3}{n + 0.4} \tag{2-102}$$

Tabulate the values of $F(t_i)$ calculated and t_i, as shown in Table 2-5.

Table 2-5 Values of cumulative failure probability

i	1	2	...	r
t_i	t_1	t_2	...	t_r
F (t_i)	F (t_1)	F (t_2)	...	F (t_r)

2) Plotting points of $[t_i, F(t_i)]$ in the t- F(t) coordinate system on a piece of Weibull probability paper on the basis of data in Table 2-5.

3) If all points plotted are basically on the same line, then the failure distribution type can be judged to be the Weibull distribution with $v = 0$.

(3) The Graphical verification method for the Weibull distribution when $v \neq 0$ is as follows: Plotting points of (X_i, Y_i) in the $X - Y$ coordinate system on a piece of Weibull probability paper, or plotting points of $[t_i, F(t_i)]$ in the t- F(t) coordinate system, if the locus is not a line but a curve, then the failure distribution type can be eliminated from the Weibull distribution with $v = 0$. However, since it might be the Weibull distribution with $v \neq 0$, it cannot be eliminated from the Weibull distribution.

1) Extend the curve obtained through points of $[t_i, \quad F(t_i)]$ plotted in the t-$F(t)$ coordinate system on the piece of Weibull probability paper and intersect with the t coordinate at the point M, the value at the point M is the estimation value $\hat{v}$ of v.

2) Calculate t_i' via $t_i' = t - \hat{v}$ and tabulate as shown in Table 2-6. Then plot points of $[t_i', \quad F(t_i)]$ in the t- $F(t)$ coordinate system on the piece of Weibull probability paper (or plot points of $[t_i, \quad F(t_i)]$ and translate a distance $\hat{v}$ to the left and obtain a group of new points as shown in Figure 2-25.)

3) If all points plotted in the previous step are basically on the same line, then the failure distribution type can be judged to be the Weibull distribution with $v \neq 0$.

Table 2-6 Values of t_i' and $F(t_i)$

i	1	2	...	r
t_i'	t_1'	t_2'	...	t_r'
F (t_i)	F (t_1)	F (t_2)	...	F (t_r)

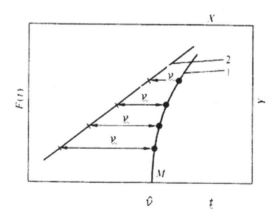

Figure 2-25 Weibull distribution with $v \neq 0$, 1-The curve obtained through points plotted on the basis of $[t_i, \quad F(t_i)]$; 2- The curve obtained through points plotted on the basis of $[t_i', \quad F(t_i)]$

For example: If there are 20 contactors of some model, the test terminated when ten contactors failed during the test, the life data are as shown in Table 2-7. Try to estimate the failure distribution type of this model of contactors and verify your estimation through graphical verification methods.

Table 2-7 Life data of some model of contactors (unit: 10^5)

6.6	7.7	8.5	9.8	10.6	11.2	12.7	13.4	14.1	14.9

Solution: draw the reliability curve firstly to estimate the failure distribution type. Calculate $R(t_i)$ and tabulate as shown in Table 2-8.

Table 2-8 $R(t_i)$、 $F(t_i)$ and t'_i values of contactors of some model

i	1	2	3	4	5	6	7	8	9	10
$R(t_i)$	0.975	0.925	0.875	0.825	0.775	0.725	0.675	0.625	0.575	0.525
$F(t_i)$	0.025	0.075	0.125	0.175	0.225	0.275	0.325	0.375	0.425	0.475
t'_i	1.3	2.4	3.2	4.5	5.3	5.9	7.4	8.1	8.8	9.6

Next, draw the reliability curve on the basis of data of t_i in Table 2-7 and $R(t_i)$ in Table 2-8 as shown in Figure 2-26. Compare the reliability curve in Figure 2-26 with the reliability curves of various failure distribution types in Table 2-2, we can see its shape is close to the reliability curve of Weibull distribution under $m>1$. So the failure distribution type is estimated to be Weibull distribution.

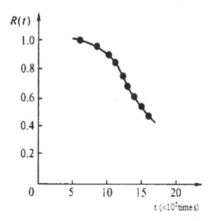

Figure 2-26   Reliability curve

Then we use a piece of Weibull probability paper to verify its failure distribution type. Calculate $F(t_i)$ on basis of $F(t_i) = 1 - R(t_i)$ and tabulate the results in Table 2-8. According to data of t_i in Table 2-7 and $F(t_i)$ in Table 2-8, plot points of $[t_i, F(t_i)]$ on the Weibull probability paper. Its locus is approximated to be a curve (as shown in curve 1 in Figure 2-27). This curve intersects with the t coordinate at the point M. The estimation value of v $\hat{v} = 5.3 \times 10^5$ according to the t value of M.

Now calculate t_i' according to $t_i' = t - \hat{v}$, and list it in Table 2-8. Plot points of $[t'_i, F(t_i)]$ in the t- F (t) coordinate system on the Weibull probability paper (as shown in Figure 2-27). From Figure 2-27 we can see the locus of these points is approximated to be a line, so the failure distribution type of this model of contactors can be considered to be Weibull distribution with $v \neq 0$.

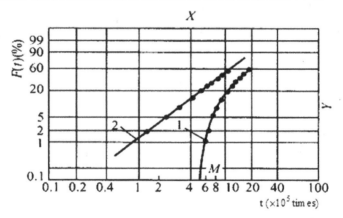

Figure 2-27 Graphical verification of Weibull distribution with $v \neq 0$
1-The curve obtained through points plotted on the basis of $[t_i, F(t_i)]$
2- The curve obtained through points plotted on the basis of $[t_i', F(t_i)]$

4.3.1.3 VERIFICATION OF WHETHER THE FAILURE DISTRIBUTION TYPE IS NORMAL DISTRIBUTION VIA NORMAL PROBABILITY PAPER

(1) Structural principles of normal probability paper- Take t as the abscissa and u as the ordinate, and add the F (t) coordinate on the ordinate on basis of the

following relations, then the schematic diagram of normal probability paper is formed as shown in Figure 2-28.

$$F(t) = \Phi(u) = \int_{-\infty}^{u} \frac{1}{\sqrt{2\pi}} e^{-\frac{x^2}{2}} dx \qquad (2\text{-}103)$$

where: $\quad u = \dfrac{t - \mu}{\sigma}$ $\qquad\qquad\qquad\qquad$ (2 101)

   To facilitate practical use, we usually translate the $F(t)$ coordinate to the left side of the probability paper and translate the t coordinate to the bottom of the probability paper (u coordinate is usually not marked), then a piece of commonly used normal probability paper is formed, as shown in Figure 2-29.

(2) Graphical verification method of normal distribution type - If all points of $[t_i,$ $F(t_i)]$ plotted on the normal probability paper is approximately on a line, then the failure distribution type can be judged to be normal distribution.

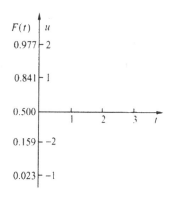
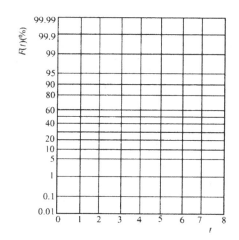

Figure 2-28

Schematic diagram of normal probability paper

Figure 2-29

Normal probability paper

4.3.2 χ^2 VERIFICATION METHOD FOR FAILURE DISTRIBUTION TYPES (IT FITS THE SITUATION WITH THE SAMPLE CAPACITY N ≥ 50).

4.3.2.1   BASIC CONCEPTS OF χ^2 VERIFICATION METHOD FOR FAILURE DISTRIBUTION TYPES

If $F_0(x)$ is the known distribution function (usually called the theoretic distribution function) and $F(x)$ is the distribution function of the population, x_1, x_2, ..., x_n is a group of observed sample values in the population with the distribution function $F(x)$. The so-called χ^2 verification method for failure distribution types is a type of method to verify the correctness of the hypothesis of H_0 for the population distribution where H_0 refers to the hypothesis that $F(x)$ is equal to $F_0(x)$, i.e.

$$H_0 \text{ is when } F(x) = F_0(x) \tag{2-105}$$

When performing the χ^2 verification, the type and parameters of the theoretic distribution function $F_0(x)$ both are known. Actually the type of $F_0(x)$ usually can be confirmed through estimation methods above mentioned for failure distribution type. However, its parameters are usually unknown. Now we should estimate its parameters then perform the χ^2 verification.

First, divide the interval $(-\infty, \infty)$ into k disjoint intervals, herein, $i=1$, 2, ..., k. their demarcation points are a_2, a_3, ..., a_k, , herein a_1 and a_{k+1} can be $-\infty$ and ∞ respectively. Widths of above intervals can be different, and the partitioning method depends on the specific situation. The number of the observed values of the sample x_1, x_2, ..., x_n in the interval i (a_i, a_{i+1}) m_i is called its actual frequency. The product of the probability that the random variable lies within the interval (a_i, a_{i+1}) and the capacity of the sample n is nP_i. It's called the theoretical frequency of the interval i. Herein, P_i can be calculated through formula (2-106) on basis of known theoretical distribution function $F_0(x)$.

$$P_i = P(a_i \leq X \leq a_{i+1}) = F_0(a_{i+1}) - F_0(a_i) \tag{2-106}$$

The statistic χ^2 is as shown in the following formula:

$$\chi^2 = \sum_{i=1}^{k} \frac{(m_i - nP_i)^2}{nP_i} \tag{2-107}$$

When the statistic $\chi^2 > c_1$ (c_1 is a constant), we can reject the above hypothesis. The range of $\chi^2 > c_1$ is usually called the rejection region; when the statistic $\chi^2 > c_1$, accept the above hypothesis.

4.3.2.2 THE DETERMINATION METHOD FOR THE VALUE OF CONSTANT C1 (THE DETERMINATION OF THE REJECTION REGION)

However, even if the hypothesis H_0 is right, because of the judgment is made on the basis of a sample, the hypothesis H_0 may still be rejected. Obviously, this is an error. The probability of committing such an error is recorded as a, usually called the significance level, i.e.

$$P(\text{refuse } H_0 \mid H_0 \text{ is true}) = P(X^2 > C1 \mid H_0 \text{ is true}) = a \tag{2-108}$$

Of course, we hope to limit the significance level under some certain value, usually take $a=0.05$ to decrease the probability of making errors. Where

$$c_1 = \chi^2_{1-a}(k - r - 1) \tag{2-109}$$

Thus, the rejection region is

$$\chi^2 > \chi^2_{1-a}(k - r - 1) \tag{2-110}$$

where $\chi^2_{1-a}(k - r - 1)$ refers to an underside quantile of χ^2 distribution of the freedom degree k-r-1. The detailed data can be found in the attachment the underside quantile table of χ^2 distribution (Appendix 2).

4.3.2.3 BASIC PROCESSES OF X2 VERIFICATION METHOD FOR FAILURE DISTRIBUTION TYPES

(1) According to the previous estimation of the failure distribution type, making an assumption about the function type of the theoretical distribution function $F_0(x)$, i.e. the failure distribution type of the population.

(2) Working out estimation values of each parameter in the theoretical distribution function $F_0(x)$.

(3) Making an hypothesis H_0 for the distribution function $F_0(x)$ of the population, i.e. H_0 refers to $F(x) = F_0(x)$.

(4) Dividing $(-\infty, \ \infty)$ into k intervals.

(5) Making a statistics about the actual frequencies m_i ($i = 1, \ 2, \ \ldots, \ k$) in each interval.

(6) Working out the probability P_i that the random variable X lies within each interval on the basis of the known theoretical distribution function $F_0(x)$.

(7) Calculating the theoretical frequency nP_i of each interval (the theoretical frequency for each interval should not be less than 5, preferable more than 10.). Otherwise, intervals should be properly merged to satisfy above requirements for the values of nP_i.

(8) Tabulate to indicate m_i, P_i, nP_i and $\dfrac{(m_i - nP_i)^2}{nP_i}$.

(9) Calculating the statistic of χ^2 on the basis of formula (2-107).

(10)   Selecting the significant level value a.

(11)   Calculating out the rejection region $\chi^2 > \chi^2_{1-a}(k - r - 1)$, where, $\chi^2_{1-a}(k - r - 1)$ can be searched in the appendix on the basis of the degree of freedom $(k - r - 1)$ and the value of 1- a.

(12)   Making a judgment: when $\chi^2 > \chi^2_{1-a}(k - r - 1)$, reject the hypothesis, when $\chi^2 < \chi^2_{1-a}(k - r - 1)$ accept the hypothesis.

For example: There are 50 relays of one model, and the life data are shown in Table 2-9.The failure distribution type of relays of this model is estimated to be normal distribution. Try to verify whether the life of relays of this model follows the normal distribution through χ^2 verification method (the significance level a= 0.05).

Table 2-9 Life data of 50 relays (unit: 10^5)

3.5	6	7.2	7.8	8.5	9	9.6	10.2	10.4	10.6
10.9	11.3	·11.9	11.9	12.5	12.7	12.8	13	13.3	13.6
13.9	14.1	14.4	14.4	14.5	14.6	14.7	14.8	15	15.1
15.3	15.6	15.8	15.9	16.3	16.4	16.5	16.7	16.8	17
17.3	17.6	17.9	18.8	19.8	20.6	21.4	21.6	21.9	25.6

Solution:

(1) Supposing the type of the theoretical distribution function $F_0(x)$ is the normal distribution $N (\mu, \ \sigma^2)$.

(2) Performing group statistics for the life data of relays of this model, frequency of each group and the medium value and so on refer to Table 2-10.

Table 2-10 Group statistics for the life data of relays of this model

i	Life range (Unit: 10^5times)	Tzi (10^5 times)	Δmi	f_i^*	Fi
1	0~4	2	1	0.02	0.02
2	4~8	6	3	0.06	0.08
3	8~12	10	10	0.2	0.28
4	12~16	14	20	0.4	0.68
5	16~20	18	11	0.22	0.90
6	20~24	22	4	0.08	0.98
7	24~28	26	1	0.02	1

(3) Working out the average life $\bar{t}$ of the sample. Take it as the estimation value of the average life μ of the population, i.e.

$$\hat{\mu} = \bar{t} = \frac{1}{50}(2 \times 1 + 6 \times 3 + 10 \times 10 + 14 \times 20 + 18 \times 11 + 22 \times 4 + 26 \times 1) \times 10^5 = 14.24 \times 10^5 \, times$$

We can obtain the life standard deviation $s = 4.79 \times 10^5$ and take it as the estimation value $\hat{\sigma}$ of the life standard deviation of the population.

(4) Making an assumption H_0 for the distribution function $F_0(x)$ of the population, i.e. H_0 refers to

$$F(t) = F_0(t) = \int_{-\infty}^{t} \frac{1}{\sqrt{2\pi} \times 4.79 \times 10^5} e^{-\frac{(x - 14.24 \times 10^5)^2}{45.89 \times 10^{10}}} \, dx$$

(5) Dividing $(-\infty, \infty)$ into 5 intervals, i.e. $(-\infty, 10 \times 10^5$ times$)$、$(10 \times 10^5$ times, 12×10^5 times$)$、$(12 \times 10^5$ times, 16×10^5 times$)$、$(16 \times 10^5$ times, 18×10^5 times$)$、$(18 \times 10^5$ times, $\infty)$。

(6) Making a statistics of the actual frequencies m_i for each interval and tabulate the results in Table 2-11.

(7) Calculating the probability that the random variable X lies within each interval. Tabulate the results in Table 2-11.

Table 2-11 Statistics table for χ^2 verification

Interval number i	Interval range (Unit: 10^5 times)	m_i	p_i	nP_i	$\dfrac{(m_i - nP_i)^2}{nP_i}$
1	$-\infty \sim 10$	7	.1881	9.41	.62
2	$10 \sim 12$	7	.1318	6.59	.03
3	$12 \sim 16$	20	.3232	16.16	.91
4	$16 \sim 18$	9	.1406	7.03	.55
5	$18 \sim \infty$	7	.2163	10.82	1.35

Next we take the probability P_2 that the random variable lies within the interval (10×10^5 times, 12×10^5 times) as an example to demonstrate the calculation method for P_i: Let $t_{p1} = 10 \times 10^5$ times we can get

$$u_{P1} = \frac{t_{p1} - \hat{\mu}}{\hat{\sigma}} = \frac{10 \times 10^5 - 14.24 \times 10^5}{4.79 \times 10^5} = -0.885$$

From the appendix 3 we can get the corresponding value of u_{P1} i.e. $p_1 = 0.1881$.
Let $t_{p2} = 12 \times 10^5$ we can get

$$u_{P2} = \frac{t_{p2} - \hat{\mu}}{\hat{\sigma}} = \frac{12 \times 10^5 - 14.24 \times 10^5}{4.79 \times 10^5} = -0.468$$

From the appendix 3 we can get the corresponding value of u_{P2} i.e. $p_2 = 0.3199$.
Thus we can obtain

$$P_2 = p_2 - p_1 = 0.3199 - 0.1881 = 0.1318$$

(8) Calculating the theoretical frequency nP_i for each interval and tabulate the results in table 2-11.

(9) Calculating $\dfrac{(m_i - nP_i)^2}{nP_i}$ for each interval and tabulate results in Table 2-11.

(10) Calculating the statistic χ^2 according to formula (2-107) on the basis of data in Table 2-11, i.e.

$$\chi^2 = \sum_{i=1}^{k} \frac{(m_i - nP_i)^2}{nP_i} = 3.46$$

(11) Selecting the significance level $a = 0.05$.

(12) Confirming the rejection region. Let $k=5$, $r=2$, $a=0.05$, so $\chi^2_{1-a}(k-r-1) = \chi^2_{0.95}(2)$, look up the appendix 2 and $x^2_{0.95}(2)$ is equal to 5.99, thus the rejection region refers to $\chi^2 > 5.99$

(13) Making judgments. Because of $\chi^2 = 3.46 < 5.99$

So we should accept the hypothesis H_0 instead of rejecting it, i.e. under the significance level $a=0.05$, the life of relays of this model is considered to follow the normal distribution.

5 ESTIMATION OF RELIABILITY CHARACTERISTIC PARAMETERS

In actual work, people usually randomly select certain number of products and perform the life test, or collect failure data on site use to obtain the observed value of the sample of product lives (for electrical apparatus, the observed value of the sample is obtained via life test.), and work out an estimation value (or an estimation interval) close to characteristic parameters of the population.

There are two methods to calculate the estimation value (or an estimation interval) of characteristic parameters of the population via the observed value of sample.

One method is to draw figure in the coordinate system on pieces of various probability paper, usually called the graphical estimation method. Its disadvantage is that the result usually varies from person with poor accuracy. However, it's very convenient, intuitive and easy to understand, and applies to the curtailed life test, so the graphical estimation method is still widely used. Especially for some situations with low requirements for accuracy, its advantages are very outstanding.

The other method is to use the mathematical analysis method for calculation, usually referred to as numerical analysis method, which can be divided into point estimate method and interval estimate method.

5.1 THE POINT ESTIMATION OF RELIABILITY CHARACTERISTIC PARAMETERS

5.1.1 CONCEPT OF THE POINT ESTIMATION

The point estimate method is to calculate the estimation value of reliability characteristic parameters on the basis of the observed value of sample.

5.1.2   THE POINT ESTIMATION OF CHARACTERISTIC PARAMETERS UNDER FULL LIFE TEST

The full life test refers to a life test which will be terminated until all test samples to failure. If n samples are selected to perform the test and let t_1, t_2, ..., t_n be life observed values of the sample, then the mean value $\bar{t}$ of observed values for sample lives can be considered as the point estimation value $\hat{\mu}$ of the average life μ of the population.

$$\hat{\mu} = \bar{t} = \frac{1}{n} \sum_{i=1}^{n} t_i \tag{2-111}$$

Similarly, the life variance and life standard deviation of the sample can be considered as the point estimation value of the life variance and life standard deviation of the population, i.e.

$$\hat{\sigma}^2 = s^2 = \frac{1}{n-1} \sum_{i=1}^{n} (t_i - \bar{t})^2 \tag{2-112}$$

$$\hat{\sigma} = s = \sqrt{\frac{1}{n-1} \sum_{i=1}^{n} (t_i - \bar{t})^2} \tag{2-113}$$

When product life follows the single-parameter exponential distribution, the point estimation value of the failure rate λ refers to

$$\hat{\lambda} = \frac{1}{\hat{\theta}} = \frac{n}{\sum_{i=1}^{n} t_i} \tag{2-114}$$

When product life follows the exponential distribution with two parameters, the point estimation value of the position parameter ν refers to

$$\hat{\nu} = t_1 - \frac{1}{n(n-1)} \left[\sum_{i=1}^{n} t_i - nt_i \right] \tag{2-115}$$

The point estimation value of the failure rate λ refers to

$$\hat{\lambda} = \frac{1}{\frac{1}{n}\sum\limits_{i=1}^{n} t_i - \hat{v}} = \frac{n}{\sum\limits_{i=1}^{n} t_i - n\hat{v}}$$ (2-116)

5.1.3   THE POINT ESTIMATION OF CHARACTERISTIC PARAMETERS UNDER CURTAILED LIFE TEST

The curtailed life test refers to the life test which will be terminated before all test samples to failure. It can be divided into time curtailed life test and failure curtailed life test on the basis of the withdrawal method of the test. Time curtailed life test means the test will be terminated when the failure number of test products has reached the specified number since the test begins.

Failure curtailed life test means the test will be terminated when the test has reached the specified cutoff time from the beginning. Time curtailed life test or failure curtailed life test can be further divided into test with replacement and test without replacement.

The life test with replacement is to ensure normal operation of the device or equipment. When any one component loses its functions, it will be replaced immediately. Thus, the total number of the component in the device or equipment remains unchanged with out regard to the number of failures during the test. The life test without replacement refers to the situation that the sample won't be replaced if any failure occurs. Instead, the rest normal samples will continue to perform the life test.

In conclusion, the life test can be divided into four types as follows on the basis of curtailed types and the replacement situation: Time curtailed life test with replacement, Time curtailed life test without replacement, failure curtailed life test with replacement, failure curtailed life test without replacement. For electrical apparatus, failure curtailed life test without replacement and time curtailed life test without replacement are mostly used.

(1) When the life follows single-parameter exponential distribution, the point estimation of reliability characteristic parameters under time curtailed life test without replacement.

If there are n samples performing the life test and terminate the test when the sample r has lost its functions, so the life data are t_1, t_2, ..., t_r , then the maximum likelihood estimation method can be used to calculate the point estimation value $\hat{\theta}$ for the average life of the product θ.

The basic principles for maximum likelihood estimation method refer to: Premise the characteristic parameters to be estimated refers to the average life θ. One point estimation value of the average life can be calculated out for observed life values of one group of samples; another point estimation value $\hat{\theta}$ can be obtained via the same formula for observed life values of another group of samples; Due to the randomness of samples, these two values $\hat{\theta}$ are usually different, so the point estimation value $\hat{\theta}$ of average life is also a random variable. We select one $\hat{\theta}$ value which owns the largest probability for the observed value of the sample from all possible values $\hat{\theta}$. This value is the maximum likelihood estimation value; its computational formula is as follows:

$$\hat{\theta} = \frac{1}{r}\left[\sum_{i=1}^{r} t_i + (n-r)t_r\right] = \frac{T}{r} \tag{2-117}$$

The point estimation value of the failure rate refers to

$$\hat{\lambda} = \frac{1}{\hat{\theta}} = \frac{r}{\left[\sum_{i=1}^{r} t_i + (n-r)t_r\right]} = \frac{r}{T} \tag{2-118}$$

where, T refers to the total test time; For failure curtailed life test, its value is as follows:

$$T = \sum_{i=1}^{r} t_i + (n-r)t_r \tag{2-119}$$

(2) When the life follows single-parameter exponential distribution, the point estimation of reliability characteristic parameters under the time curtailed life test without replacement.

If there are n samples performing the life test and terminate the test when it reaches the specified time, so the life data are t_1, t_2, ..., t_r, then the maximum likelihood method can be used to calculate the point estimation value $\hat{\theta}$ for the average life θ of the product. Its computational formula is as follows:
The point estimation value of average life refers to:

$$\hat{\theta} = \frac{1}{r}\left[\sum_{i=1}^{r} t_i + (n-r)t_c\right] = \frac{T}{r} \tag{2-120}$$

The point estimation value of the failure rate refers to

$$\hat{\lambda} = \frac{r}{\left[\sum\limits_{i=1}^{r} t_i + (n-r)t_c\right]} = \frac{r}{T} \tag{2-121}$$

where, T is the total test time; For failure curtailed life test, its value is as follows:

$$T = \sum_{i=1}^{r} t_i + (n-r)t_c \tag{2-122}$$

(3) When the life follows two-parameter exponential distribution, the point estimation of average life under time curtailed life test without replacement.

Premise there are n test samples performing the life test without replacement and terminate the test when the test time $t = t_c$, their life data are t_1, t_2, ..., t_r, considering the transformation of data, the point estimation value of average life when the life follows two-parameters exponential distribution can be calculated out via formula (2-123), i.e.

$$\hat{\theta} = \frac{T}{r} - \left(\frac{n}{r} - 1\right)\hat{v} \tag{2-123}$$

where $\quad \hat{v} = t_1 - \frac{1}{n(r-1)}\left[\sum_{i=1}^{r} t_i + (n-r)t_r - nt_1\right] \tag{2-124}$

In the formula, T should be calculated out via formula (2-122).

(4) When the life follows two-parameters exponential distribution, the point estimation of average life under failure curtailed life test without replacement.

The computational formula for the point value $\hat{\theta}$ of average life is the same as formula (2-123); however, the value of T in the formula should be calculated through formula (2-119).

5.1.4 THE POINT ESTIMATION OF RELIABILITY CHARACTERISTIC PARAMETERS UNDER SMALL SAMPLES

The so-called point estimation of reliability characteristic parameters refers to the point estimation of characteristic parameters under the situation with very few failure test samples. Reliability characteristic parameters under the small samples are estimated on the basis of the amendment model for maximum likelihood estimation method (*MMLE*).The method to establish the amendment model for maximum likelihood estimation is as follows:

Premise there are n samples performing failure curtailed test and terminate the test at the t_c moment. Premise there are r samples observed that has lost their functions and the failure data are as follows: $t_1 \le t_2 \le \cdots \le t_r(t_r \le t_c)$. If the product life follows single-parameter exponential distribution, then the estimation formula for average life is as follows:

$$\hat{\theta}_M = \frac{\sum_{i=1}^{r} t_i + (n-r)t_c}{r + L} \qquad (2\text{-}125)$$

where L should be calculated through the following formula:

$$L = \begin{cases} 0, & r = n \\ \min\left\{1, \dfrac{\ln t_c - \ln t_r}{EZ_{n,r+1} - EZ_{n,r}}\right\}, & r < n \end{cases} \qquad (2\text{-}126)$$

In the above formula, the value of $EZ_{n,k}$ should be calculated through the following formula, i.e.

$$EZ_{n,k} = \ln\left[-\ln(1 - \frac{k - 0.5}{n + 0.25})\right] \qquad (2\text{-}127)$$

5.1.5   POINT ESTIMATION OF RELIABILITY CHARACTERISTIC PARAMETERS FOR ZERO-FAILURE DATA

For expensive electrical apparatus that have long life and various types of test items, such as breakers, switchgear and so on, it is expensive to perform the test for each regularly. To reduce its test cost, a life test method with different curtailed time can be employed when assessing their reliability performances, i.e. simultaneously perform the test for test products and inspect some products among the population from time to time, terminate the test for inspected test product regardless of failures. Thus, the test cost can be greatly reduced.

5.1.5.1   BASIC METHODS

The situation with no failure product observed may occur when performing the test on the basis of the above test method, e.g. zero-failure data, i.e. there are m_i

products with no failure at the t_i moment, i.e. i=1,2,3,…, k, $t_1 \le t_2 \le \cdots \le t_k$, $n = m_1 + m_2 + \ldots + m_k$, $n_i = m_i + m_{i+1} + \ldots + m_k$, such type of data is called zero-failure data in this book. For such type of data, major errors may be produced when performing the estimation through classic methods. Grading *Bayes* method is employed in this book when performing the estimation for reliability characteristic parameters with zero failure data for electrical apparatus. Take Weibull distribution as an example to demonstrate it.

5.1.5.2 THE ESTIMATION OF RELIABILITY CHARACTERISTIC PARAMETERS

Premise the product life follows Weibull distribution with two parameters, and its cumulative failure distribution function refers to $F(t) = 1 - e^{-\left(\frac{t}{\eta}\right)^m}$, herein, η and m refers to unknown parameters. Premise the cumulative failure probability is p_i under $t = t_i$, then $p_i = 1 - e^{-\left(\frac{t_i}{\eta}\right)^m}$, i=1, 2… k. Replace p_i with $\hat{p}_i$ and employ weighed least square method, take the weight factor as $\omega_i = m_i t_i / \sum_{i=1}^{k} m_i t_i$, then the estimation of η and m is:

$$\hat{\eta} = e^{(BC - AD)/(B - A^2)} \tag{2-128}$$

$$\hat{m} = (B - A^2)/(D - AC) \tag{2-129}$$

where, $$A = \sum_{i=1}^{k} \omega_i \ln \ln(1 - \hat{p}_i)^{-1} \tag{2-130}$$

$$B = \sum_{i=1}^{k} \omega_i (\ln \ln(1 - \hat{p}_i)^{-1})^2 \tag{2-131}$$

$$C = \sum_{i=1}^{k} \omega_i \ln t_i \tag{2-132}$$

$$D = \sum_{i=1}^{k} \omega_i \ln t_i \cdot \ln \ln(1 - \hat{p}_i)^{-1} \tag{2-133}$$

Then the estimation of product reliability under $t = \tau$ is

$$R(\tau) = e^{-\left(\frac{\tau}{\hat{\eta}}\right)^{\hat{m}}}$$
(2-134)

Other reliability characteristic parameters can be obtained through the relations between reliability characteristic parameters and distribution parameters.

For example, a type of electrical apparatus whose life follows the Weibull distribution with two parameters, to evaluate its reliability indices we randomly select 10 samples to perform the test. When the operating cycles have reached 0.8×10^5, inspect two of the test samples. There are no failures and tests for these two samples are terminated. When the operating cycles have reached 1×10^5 inspect three of the rest test samples. There are no failures and tests for these three samples are terminated. When the operating cycles have reached 1.8×10^5 inspect three of the rest test samples. There are no failures and tests for these three samples are terminated. When the operating cycles have reached 1.2×10^6 inspect the last two samples. There are no failures too. The test data are as shown in Table 2-12. Try to figure out the estimation value $\hat{\mu}$ of the average life for this electrical apparatus.

Table 2-12 Test data

i	t_i(times)	m_i	n_i
1	80000	2	10
2	100000	3	8
3	180000	3	5
4	1200000	2	2

Solution:
(1) ascertain the priori information - Premise there are N samples of such model performing the test previously and the life test data are as follows t_1, t_2, ..., t_N, perform the point estimation for the previous reliability of such type of products. The reliability of such product under the specified time t_L can be calculated out.

$$R(t_L) = \frac{\sum_{i=1}^{N} V_i}{N}$$
(2-135)

$$\text{where, } V_i = \begin{cases} 1 & t_i \geq t_L \\ 0 & t_i < t_L \end{cases} \tag{2-136}$$

For example, the reliability can be estimated to be 0.9 under the operating cycles $t_L = 10^6$, i.e. $R(10^6 \text{times}) = 0.9$. This empirical data can be used as the priori information for the estimation of reliability characteristic parameters for this batch of products.

(2) Ascertain the parameter c = According to the above mentioned priori information, we can calculated out through formula (2-128)~(2-134) when $c=80\sim90$, estimation values of position parameter m, true scale parameter η are $\hat{m}$ and $\hat{\eta}$, and when the life is equal to 10^6 times, the estimation value of product reliability $\hat{R}$ (1×10^6 times) and the estimation value $\hat{\mu}$ ($\hat{\mu} = \eta\Gamma\left(1+\dfrac{1}{m}\right)$) of average life μ are as shown in Table 2-13. The reliability R (10^6 times) is equal to 0.9, value in the priori information which mostly approximate c is equal to 83. Thus determine the uniform distribution upper limit c of the hyper parameter b is 83.

(3) Estimate the average life $\hat{\mu}$ of the product

Thus, the product life follows Weibull distribution with two parameters under the position parameter $m = 0.44$ and the true scale parameter $\eta = 1.67 \times 10^8$ times. Its point estimation value for the average life refers to $\hat{\mu} = 4.22 \times 10^8$ times.

Table 2-13 Calculation results

c	$\hat{m}$	$\hat{\eta}$ (10^8 times)	$\hat{\mu}$ (10^8 times)	$\hat{R}$ (10^6 times)
80	0.441	1.57	3.98	0.8978
81	0.440	1.60	4.06	0.8986
82	0.440	1.64	4.14	0.8994
83	0.440	1.67	4.22	0.9001
84	0.440	1.70	4.30	0.9009
85	0.439	1.74	4.39	0.9016
86	0.439	1.77	4.47	0.9023
87	0.439	1.81	4.55	0.9031
88	0.439	1.84	4.64	0.9038
89	0.439	1.88	4.72	0.9045

5.2 THE INTERVAL ESTIMATION OF RELIABILITY CHARACTERISTIC PARAMETERS

5.2.1 CONCEPTS OF INTERVAL ESTIMATION

The confidence interval estimation method(interval estimation method for short) refers to the method to give an estimation interval for one characteristic parameters of the population. Detailed processes for interval estimation are as follows:

If one characteristic parameters Θ of reliability of the-population needs to be estimated, then an interval (Θ_L, Θ_U) should be calculated out via certain processes. The probability that the interval covers the true value of Θ is usually referred to as $1 - a$, its mathematical expression is as follows:

$$(\Theta_L \leq \Theta \leq \Theta_U) = 1-a \tag{2-137}$$

$1-a$ is usually called the confidence degree or the confidence level, while a is referred to as the significance level. The interval (Θ_L, Θ_U) is called confidence interval with Θ_L as its confidence lower limit and Θ_U as its confidence upper limit. Because of such interval estimation method should figure out the upper and lower confidence limit, so it's also called the interval estimation method via the calculation of bilateral confidence limits.

The confidence degree should be specified firstly when performing the interval estimation. The confidence degree $1-a$ is usually 0.9 or 0.6.

5.2.2 THE INTERVAL ESTIMATION WITH BILATERAL CONFIDENCE LIMITS OF RELIABILITY CHARACTERISTIC PARAMETERS UNDER EXPONENTIAL DISTRIBUTION

5.2.2.1 THE INTERVAL ESTIMATION FOR AVERAGE LIFE UNDER TIME CURTAILED LIFE TEST WITHOUT REPLACEMENT

Through mathematical deduction we can get:

$$P\left(\frac{2T}{\chi^2_{1-\frac{a}{2}}(2r)} \leq \theta \leq \frac{2T}{\chi^2_{\frac{a}{2}}(2r)} \right) = 1-a \tag{2-138}$$

Thus the lower confidence limit θ_L and the upper confidence limit θ_U for average life under time curtailed life test without replacement are respectively

$$\theta_L = \frac{2T}{\chi^2_{1-\frac{a}{2}}(2r)}$$

(2-139)

$$\theta_U = \frac{2T}{\chi^2_{\frac{a}{2}}(2r)}$$

(2-140)

where, $\chi^2_{1-\frac{a}{2}}(2r)$ and $\chi^2_{\frac{a}{2}}(2r)$ refer to the underside quantile of the χ^2

distribution with freedom degrees 2r.

5.2.2.2   THE INTERVAL ESTIMATION FOR AVERAGE LIFE UNDER FAILURE CURTAILED LIFE TEST WITHOUT REPLACEMENT

It can be proved that the lower confidence limit θ_L and the upper confidence limit θ_U for average life under time curtailed life test without replacement are respectively

$$\theta_L = \frac{2T}{\chi^2_{1-\frac{a}{2}}(2r+2)}$$

(2-141)

$$\theta_U = \frac{2T}{\chi^2_{\frac{a}{2}}(2r)}$$

(2-142)

where, $\chi^2_{1-\frac{a}{2}}(2r+2)$ refers to the underside quantile of the χ^2 distribution with

freedom degrees 2r.

5.2.3   THE ESTIMATION OF THE LOWER LIMIT FOR CHARACTERISTIC PARAMETERS UNDER EXPONENTIAL DISTRIBUTION WITH UNILATERAL CONFIDENCE LIMIT

The true value θ of average life is guaranteed to exceed a certain value θ'_L with the confidence degree of 1- a, i.e.

$$P(\theta_L' \le \theta < \infty) = 1 - a \qquad (2\text{-}143)$$

Such interval estimation method is usually referred to as the interval estimation method for average life with unilateral confidence limit.
Through mathematical deduction we can get:

$$P\left[\frac{2T}{\chi_{1-a}^2(2r)} \le \theta < \infty\right] = 1 - a \qquad (2\text{-}144)$$

For the product life which follows the distribution with single parameter, the lower confidence limit θ_L' in the lower limit estimation method for the average life under time curtailed life test without replacement is

$$\theta_L' = \frac{2T}{\chi_{1-a}^2(2r)} \qquad (2\text{-}145)$$

For failure curtailed life test without replacement, the lower confidence limit in the lower limit estimation method for the average life is

$$\theta_L' = \frac{2T}{\chi_{1-a}^2(2r+2)} \qquad (2\text{-}146)$$

5.3 THE GRAPHICAL ESTIMATION OF RELIABILITY CHARACTERISTIC PARAMETERS

5.3.1 THE GRAPHICAL ESTIMATION OF RELIABILITY CHARACTERISTIC PARAMETERS FOR EXPONENTIAL DISTRIBUTION

5.3.1.1 THE GRAPHICAL ESTIMATION OF THE SINGLE-PARAMETER EXPONENTIAL DISTRIBUTION.

(1) The graphical estimation of the parameter λ

Find A point with the indication of 0.368 on the $R(t)$ coordinate on a piece of unilateral logarithm graph paper. Draw a horizontal line from A point towards the right, which intersects with the regression line at the point B. The t coordinate of the intersection point B is the estimation value $\hat{\theta}$ of the average life θ, as shown in Figure 2-30.

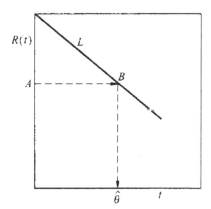

Figure 2-30 The graphical estimation for the parameter λ

(2) The graphical estimation for reliability characteristic parameters

The graphical estimation for the reliability function $R(t)$: Find the specified operating time on t coordinate in Figure 2-30, and draw a line perpendicular to the regression straight line L. The $R(t)$ coordinate of the intersection point between this vertical line and the regression straight line refers to the graphical estimation $\hat{R}(t)$ of the reliability function $R(t)$.

The graphical estimation of the cumulative failure probability $F(t)$ is:

$$\hat{F}(t) = 1 - \hat{R}(t).$$

Find the point with the indication of R on the $R(t)$ coordinate and draw a horizontal line, the intersection between the horizontal line and the regression straight line L refers to the estimation value $\hat{t}_R$ of the reliability life t_R.

5.3.1.2   THE GRAPHICAL ESTIMATION FOR THE TWO-PARAMETERS EXPONENTIAL DISTRIBUTION

(1) The graphical estimation of the position parameter ν

Find D point with the indication of 1 on the $R(t)$ coordinate on a piece of unilateral logarithm graph paper, draw a horizontal line from D point towards the right, which intersects with the regression line at the point E. The t coordinate of the intersection point E refers to the estimation value $\hat{\nu}$ of the position parameter ν, as shown in Figure 2-31.

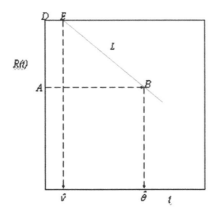

Figure 2-31 Graphical estimations for the position parameter v and the parameter λ

(2) The graphical estimation of the parameter λ

Find A point with the indication of 0.368 on the $R(t)$ coordinate on a piece of unilateral logarithm graph paper, draw a horizontal line from A point towards the right, which intersects with the regression line at the point B. The t coordinate of the intersection point B is the estimation value $\hat{\theta}$ of the average life θ, as shown in Figure 2-31.

The graphical estimation for the distribution parameter refers to: $\hat{\lambda} = \dfrac{1}{\hat{\theta} - \hat{v}}$

(3) The graphical estimation for reliability characteristic parameters

The graphical estimation for the reliability function $R(t)$: Find the specified operating time on t coordinate in Figure 2-31, and draw a line perpendicular to the regression straight line L. The $R(t)$ coordinate of the intersection point between this vertical line and the regression straight line refers to the graphical estimation $\hat{R}(t)$ of the reliability function $R(t)$.

The graphical estimation of the cumulative failure probability $F(t)$ is:

$$\hat{F}(t) = 1 - \hat{R}(t)$$

The graphical estimation of life standard deviation $\hat{\sigma} = \hat{\theta} - \hat{v}$.

Find the point with the indication of R on the $R(t)$ coordinate and draw a horizontal line, the intersection point between the horizontal line and the regression straight line L refers to the estimation value $\hat{t}_R$ of the reliability life t_R.

5.3.2   THE GRAPHICAL ESTIMATION OF RELIABILITY CHARACTERISTIC PARAMETERS UNDER WEIBULL DISTRIBUTION

5.3.2.1   THE GRAPHICAL ESTIMATION OF WEIBULL DISTRIBUTION UNDER $v = 0$

5.3.2.1.1   THE GRAPHICAL ESTIMATION OF WEIBULL DISTRIBUTION PARAMETERS M, T0 AND η

(1)   The estimation of the shape parameter m

Draw a line from the estimation point of m $(X=1, Y=0)$ parallel to the regression straight line $Y = mX - B$. The ordinate of the intersection point A between the parallel line and the Y coordinate refers to $-m$. Thus, the estimation value $\hat{m}$ of the shape parameter m is the absolute value of the indication of point B as shown in Figure 2-32.

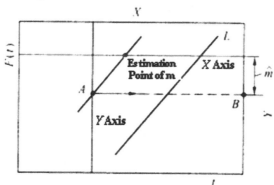

Figure 2-32   Graphical estimation method for $\hat{m}$

(2)   The estimation of the true scale parameter η

The estimation value of η can be obtained directly through Weibull probability paper: Draw a vertical line downward from the intersection point P between the regression straight line and the X coordinate, this vertical line intersects with t coordinate at the Q point. The indication of Q point refers to $\hat{\eta}$ as shown in Figure 2-33.

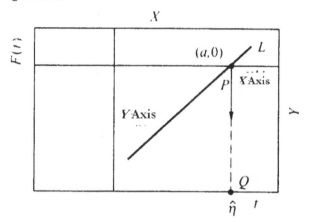

Figure 2-33 Graphical estimation method for $\hat{\eta}$

5.3.2.1.2   THE GRAPHICAL ESTIMATION OF RELIABILITY CHARACTERISTIC PARAMETERS

(1)   The graphical estimation of average life

According to the calculated $\hat{m}$ value, find out the $\hat{\mu}/\eta$ on the μ/η coordinate, then the estimation value $\hat{\mu}$ of average life μ can be calculated through formula (2-147):

$$\hat{\mu} = \hat{\eta}\frac{\hat{\mu}}{\eta} \qquad\qquad (2\text{-}147)$$

In addition, according to the calculated $\hat{m}$ value, we can find out the corresponding $F(\hat{\mu})$ on the $F(\mu)$ coordinate, and then find the point E with the indication of $F(\hat{\mu})$. Draw a horizontal line towards the right and intersect with regression straight line L at the point F, next draw a vertical line from the F point and intersect with the t coordinate at the point G, the indication of which refers to $\hat{\mu}$ (as shown in Figure 2-34).

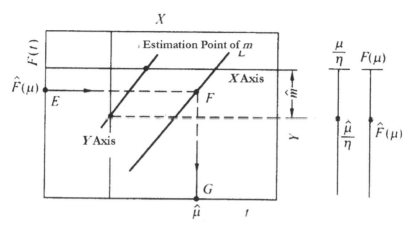

Figure 2-34 Graphical estimation method for $\hat{\mu}$.

(2) The graphical estimation of the life standard deviation σ

According to the calculated $\hat{m}$ value, find out the corresponding $\hat{\sigma}/\eta$ on the σ/η coordinate, then the estimation value for the life standard deviation can be calculated through formula (2-148), i.e.

$$\hat{\sigma} = \hat{\eta}\frac{\hat{\sigma}}{\eta} \tag{2-148}$$

In addition, according to the calculated $\hat{m}$ value, we can find out the corresponding $F(\hat{\sigma})$ on the $F(\sigma)$ coordinate, and then find the point R with the indication of $F(\hat{\sigma})$. Draw a horizontal line from the R point towards the right and intersect with the regression straight line L at the point S, then draw a vertical line downwards from the S point and intersect with the t coordinate at the point G, the indication of T point refers to the estimation value $\hat{\sigma}$ of σ (as shown in Figure 2-35).

When the shape parameter $m < 1.5$, the accuracy of μ/η , σ/η coordinate and m coordinate on the right side of Weibull probability paper is poor, thus the enlarged μ/η , σ/η coordinate and m coordinate are carved on the top side of Weibull probability paper. Thus the μ/η , σ/η coordinate and m coordinate is carved on the right side of a complete Weibull probability paper, while an enlarged the μ/η , σ/η coordinate and m coordinate is carved on the top side as shown in Figure 2-36.

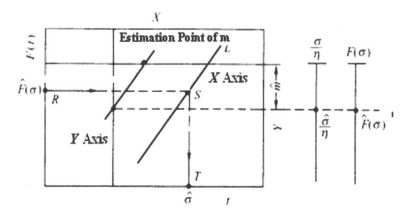

Figure 2-35 Graphical estimation method for $\hat{\sigma}$

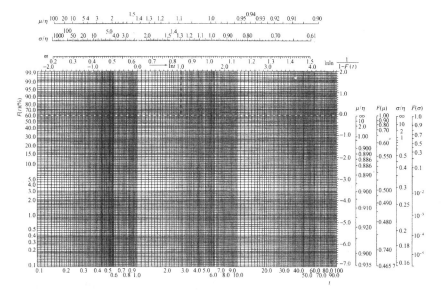

Figure 2-36 A complete Weibull probability paper

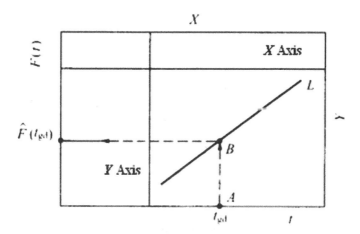

Figure 2-37 Graphical estimation method for $\hat{F}(t_{gd})$

(3)   The graphical estimation of the reliability $R(t_{gd})$ under the specified time t_{gd}

Find the A point with the indication of t_{gd} on t coordinate, draw a vertical line upwards from the point A and intersect with regression straight line L at the point B, draw a horizontal line from B point towards the left, the indication of the intersection point between this horizontal line with the F(t) coordinate should be equal to the estimation value $\hat{F}(t_{gd})$ of $F(t_{gd})$ as shown in Figure 2-37. Thus, the estimation value $\hat{R}(t_{gd})$ of $R(t_{gd})$, can be calculated through the following formula: $\hat{R}(t_{gd}) = 1 - \hat{F}(t_{gd})$

(4)   The graphical estimation of the reliability life t_R under specified reliability R

Find the C point with the indication of 1-R on F(t) coordinate, draw a horizontal line towards the right from C point and intersect with the regression straight line L at the point D, then draw a vertical line downwards from D point and the indication of the intersection point E between this vertical line and the t coordinate refers to the estimation value $\hat{t}_R$ of t_R, as shown in Figure 2-38.

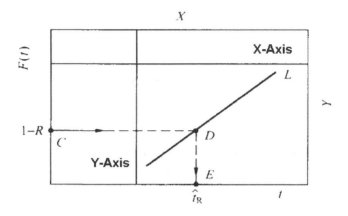

Figure 2-38 Graphical estimation method for $\hat{t}_R$

5.3.2.2 THE GRAPHICAL ESTIMATION OF WEIBULL DISTRIBUTION UNDER $v \neq 0$

5.3.2.2.1 THE GRAPHICAL ESTIMATION OF PARAMETERS *M*, T_0 AND η OF WEIBULL DISTRIBUTION

When the product life follows Weibull distribution under $v \neq 0$, plot points of $[t_i, \ F(t_i)]$ in the t- $F(t)$ coordinate system on a piece of Weibull probability paper, their locus is a curve. The indication of the intersection point between this curve and the t coordinate refers to the estimation value $\hat{v}$ of the position parameter v. The straight line linearized from this curve is also referred to as the regression straight line. On the basis of this regression straight line, using the same method employed in Weibull distribution under $v = 0$, we can obtain the estimation value of the parameters m and η.

5.3.2.2.2 THE GRAPHICAL ESTIMATION OF RELIABILITY CHARACTERISTIC PARAMETERS

(1) The graphical estimation of for average life μ

Use the μ / η coordinate and $F(\mu)$ coordinate to calculate $\hat{\mu}'$ through the same method employed under $v = 0$ (as shown in Figure 2-39).

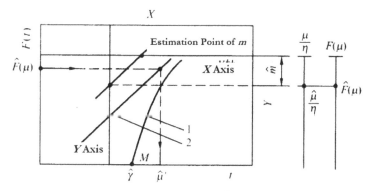

Figure 2-39   Graphical estimation method for $\hat{\mu}$ under $v \neq 0$

1- The locus of points plotted on the basis of $[(t_i,\ F(t_i))$

2- The regression straight line after linearization

The estimation value for average life should be calculated through the following formula:

$$\hat{\mu} = \hat{\mu}' + \hat{v} \tag{2-149}$$

(2)   The graphical estimation of life standard deviation σ

It is completely the same with the method employed under $v = 0$.

(3)   The graphical estimation of the reliability $R(t_{gd})$ under the specified time

t_{gd}

The process is similar to the method employed under $v = 0$. The only difference is: when drawing the vertical line from the A point with the indication equal to t_{gd} on the t coordinate, we should select the intersection point B between this vertical line and the locus of points plotted on the basis of points $[t_i,\ F(t_i)]$, and draw a horizontal line from B point towards the left the indication of the intersection point between it and the coordinate $F(t)$ refers to $\hat{F}(t_{gd})$, as shown in Figure 2-40. The estimation value $\hat{R}(t_{gd})$ of $R(t_{gd})$ to be calculated can be obtained via $\hat{R}(t_{gd}) = 1 - \hat{F}(t_{gd})$.

(4)   The graphical estimation of the reliability life t_R under specified reliability R

Find the C point with the indication equal to $1 - R$ on F (t) coordinate, draw a horizontal line towards the right from C point and intersect with the locus of points plotted on the basis of points $[t_i,\ F(t_i)]$ at the point D, then draw a vertical line downwards from D point and the indication of the intersection point E between this vertical line and the t coordinate refers to the estimation value $\hat{t}_R$ of t_R, as shown in Figure 2-41.

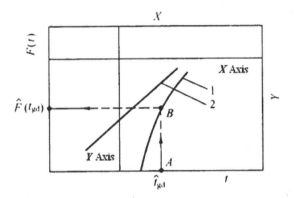

Figure 2-40 Graphical estimation method for $\hat{F}(t_{gd})$ under $v \neq 0$; 1-The locus of points plotted on the basis of $[(t_i, F(t_i)]$; 2-The regression straight line after linearization

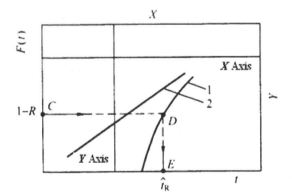

Figure 2-41 Graphical estimation method for $\hat{t}_R$ when $v \neq 0$; 1-The locus of points plotted on the basis of $[(t_i, F(t_i)]$; 2-The regression straight line after linearization

3

BASIC THEORY FOR ELECTRICAL APPARATUS RELIABILITY

1 THEORY AND METHODS OF SAMPLING FOR ELECTRICAL APPARATUS RELIABILITY

1.1 CLASSIFICATION OF SAMPLING INSPECTION PLAN

Sampling inspection plans can be classified according to the properties, uses, sampling times and operating mode, etc.

1.1.1 CLASSIFICATION ACCORDING TO THE PROPERTIES

According to the properties, sampling inspections can be classified into sampling inspections by attributes and sampling inspections by variables.

(1) Sampling Inspections by Attributes

Sample a certain number of products from a batch for inspection, classify the samples into effective and defective products according to the results, and then compare the number of the inspected defective products with the preset "acceptance number" to judge whether the batch of products is qualified or not.

(2) Sampling Inspections by Variables

Sample a certain number of products from a batch, measure certain characteristic quantity of each sample and calculate this characteristic value by statistics, and then compare it with the specified standard value to judge whether the batch of products is qualified or not.

1.1.2   CLASSIFICATION ACCORDING TO THE USES

According to the uses, sampling inspections can be classified into quality sampling inspection and reliability sampling inspection.

(1)   Quality Sampling Inspection

The sampling inspection is performed in order to inspect the product quality, such as the sampling inspection for parts and components during the production process and the sampling inspection for periodical test and delivery test of products.

(2) Reliability Sampling Inspection

The sampling inspection is performed in order to inspect the product reliability, such as the sampling inspection for the failure rate, the mean life, and the reliable life.

1.1.3   CLASSIFICATION ACCORDING TO SAMPLING TIMES

According to the sampling times, the sampling inspections can be classified into single sampling inspection, double sampling inspection, multiple sampling inspection and sequential sampling inspection.

(1)   Single Sampling Inspection

Sample a certain number of products only once from one inspected batch, and judge whether the batch of products is qualified or not according to the sampling inspection result.

(2)   Double Sampling Inspection

The batch of product can be judged as effective defective, and uncertain by the first sample inspection. In the case of uncertain it's required to sample for the second time to judge whether the batch of products is qualified or not by the first and second sampling inspection results.

(3)   Multiple Sampling Inspection

If the decisions whether to qualify the batch of products or not cannot be made according to the first and second sampling inspection results, further sampling inspections must continue until the decision can be made according to all the sampling inspection results.

(4)   Sequential Sampling Inspection

Judge whether the batch of products is qualified or not according to the sequential accumulative data from the sampling inspection results.

A brief description is given below in the case of sequential sampling inspection by attributes: take the accumulative number of samples *n* as abscissa, and the number of defectives *r* found in the *n* samples as ordinate, and draw two straight lines on the coordinate paper in some way, as shown in Figure 3-1. In Figure 3-1, Line 1 is called reject line, and Line 2 is called accept line; the area above Line 1 is called reject area, the area below Line 2 is called accept area, and the area between Lines 1 and 2 is called continuous test area.

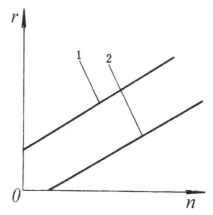

Figure 3-1 Sequential sampling inspection: 1- Reject Line   2- Accept Line

When the data point (*n*, *r*) drawn in Figure 3-1 according to the sampling inspection result is located on the accept line or within the accept area, the batch of products is judged accept. When the data point (*n*, *r*) is located on the reject line or within the reject area, the batch of products is judged reject. When the data point (*n*, *r*) is within continuous test area, the further sampling inspection is required until the decision can be made.

The characteristic of sequential sampling inspection is that it can minimize the average sample number ASN, so it's especially suitable for the inspection with high cost.

1.1.4   CLASSIFICATION ACCORDING TO THE OPERATING MODE

According to the operating mode, the sampling inspections can be classified into standard sampling inspection and adjusting sampling inspection.

(1)   Standard Sampling Inspection

Standard sampling inspection refers to the method to draft the sampling inspection scheme of the product without considering the previous sampling inspection results of the product.

(2) Adjusting Sampling Inspection

Adjusting sampling inspection refers to the method that adjusting the stringency of sampling inspection at any moment according to the change of product quality, i.e. according to the previous sampling inspection results of several batches of products. It's generally composed of three kinds of sampling inspections with different stringency: normal inspection, tightened inspection and loose inspection. Shifting from one inspection to another should comply with the predetermined shift rule. For example, using normal inspection in case of normal product quality, which can be shifted to tightened inspection in case of deteriorating product quality (i.e. at occurrence of certain reject batches in continuous batches of products); the tightened inspection can be shifted back to normal inspection in case of recovered product quality; the normal inspection can be shifted to loose inspection in case of good product quality (i.e. in the conditions that continuous batches are accepted, etc.); the loose inspection should be shifted back to normal inspection when the product quality declines again (i.e. occur a lot of defective products and other situation).

1.2 BASIC THEORY OF SAMPLING INSPECTION

1.2.1 ACCEPTABLE PROBABILITY OF SAMPLING INSPECTION SCHEME

The acceptable probability of sampling inspection scheme refers to the probability that a batch of products is accepted by the sampling inspection scheme. It's obvious that acceptable probability is related to the actual percent defective p of the batch of products, denoted by $L(p)$.

(1) Acceptable Probability of Single Sampling Inspection by Attributes

The so-called single sampling inspection by attributes refers to sampling n samples randomly from one batch of products which the total number is N, and judging the batch of products is acceptable if the number of inspected failures $r \leq A_c$ (A_c is the acceptance number); judging the batch of products is rejected if $r > A_c$. The block diagram of single sampling inspection by attributes is shown in Figure 3-2.

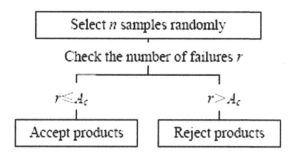

Figure 3-2 Block diagram of single sampling inspection by attributes

The acceptable probability equals to the sum of probabilities when the number of failures r of the n random samples are respectively 0, 1, ..., A_c. In case of larger N ($N>10n$), this probability can be calculated by binomial probability, i.e.

$$L(p) = \sum_{r=0}^{A_c} P(r,n \mid p) = \sum_{r=0}^{A_c} C_n^r p^r q^{n-r} \qquad (3\text{-}1)$$

where q is quality rate of products.

(2) Acceptable Probability of Double Sampling Inspection by Attributes

The block diagram of typical double sampling inspection by attributes is shown in Figure 3-3.

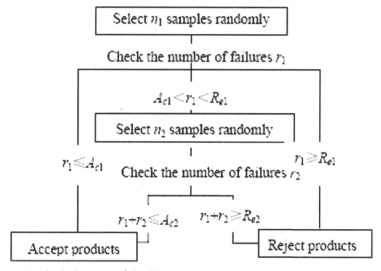

Figure 3-3 Block diagram of double sampling inspection by attributes

In Figure 3-3, A_{c1} is the primary acceptance number, R_{e1} is the primary reject number, A_{c2} is the secondary acceptance number, and R_{e2} is the secondary reject number.

The procedures of double sampling inspection scheme: sample samples n_1 randomly from one batch of products, and judge the batch of products is qualified if the number of failures r_1 is equal to or less than the primary acceptance number A_{c1}; judge the batch of products is unqualified if r_1 is equal to or larger than the primary reject number R_{e1}; judgment can not be made if r_1 is larger than A_{c1} but less than R_{e1}, so it's required to sample secondary samples n_2 for inspection; judge the batch of products is qualified if the sum of the number of failures r_2 and the number of failures r_1 of the first samples is less than or equal to the secondary acceptance number A_{c2}; judge the batch of products is unqualified if r_1+r_2 is equal to or larger than the secondary reject number R_{e2}.

The acceptable probability of double sampling inspection by attributes can be calculated by the following formula:

$$L(p) = P(r_1 \le A_{C1}) + P(A_{C1} < r_1 < R_{e1}, r_1 + r_2 \le A_{C2}) \qquad (3\text{-}2)$$

When the total number of this batch of products N is large ($N>10n$), the Formula (3-2) can be:

$$L(p) = \sum_{r_1=0}^{A_{C1}} P(r_1, n_1 \mid p) + \sum_{r_1=A_{C1}+1}^{R_{e1}-1} P(r_1, n_1 \mid p) \sum_{r_2=0}^{A_{C2}-r_1} P(r_2, n_2 \mid p)$$

$$= \sum_{r_1=0}^{A_{C1}} C_{n_1}^{r_1} p^{r_1} q^{n_1-r_1} + \sum_{r_1=A_{C1}+1}^{R_{e1}-1} C_{n_1}^{r_1} p^{r_1} q^{n_1-r_1} \sum_{r_2=0}^{A_{C2}-r_1} C_{n_2}^{r_2} p^{r_2} q^{n_2-r_2} \quad (3\text{-}3)$$

1.2.2 OPERATING CHARACTERISTIC CURVE AND PARAMETERS P_0, P_1, α AND β OF SAMPLING INSPECTION

The operating characteristic curve of a sampling inspection scheme refers to the relation of the acceptable probability $L(p)$ and percent defective p, generally called OC-curve for short.

(1) Ideal OC-curve

Ideal sampling inspection scheme: first specify a tolerable percent defective p_y, and judge that the batch of products is qualified when the actual percent defective p is less than or equal to p_y, i.e. its acceptable probability $L(p)$ should be equal to 1; judge that the batch of products is unqualified when the actual percent defective p is larger than p_y, i.e. the acceptable probability $L(p)$ should be equal to zero. So the OC-curve of ideal sampling inspection scheme should be step, as shown in Figure 3-4. However, this kind of ideal sampling inspection scheme does not exist, and the typical OC-curve is shown in Figure 3-5.

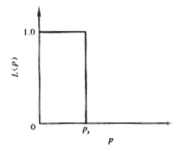

Figure 3-4 Ideal OC-curve

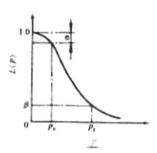

Figure 3-5 Typical OC-curve

(2) OC-curve and Parameters p_0, p_1, α and β of Actual Sampling Inspection
We generally specify two percent defective (p_0 and p_1), and the batch of products is considered acceptable when the actual percent defective p is less than or equal to p_0, so p_0 is called acceptable quality level, or AQL for short. It can be seen from Figure 3-5 that the acceptable probability $L(p_0)=1-\alpha$ at some value p_0 of p, the rejection probability is α, and α is called producer's rick; when the actual product percent defective p is larger than p_1, it should be considered the lot of products is failure, so p_1 is called lot tolerance percent defective, or LTPD for short. In some documents [such as Chinese National Standard GB/T2829 *Sampling Procedures and Tables for Periodic Inspection by Attributes (Apply to Inspection of Process Stability)*], p_1 is called unacceptable quality level. It can be seen from Figure 3-5 that when $p=p_1$, the acceptable probability is $L(p_1)=\beta$, β is called consumer's risk. If $p=p_1$, the batch of products should has been judged unqualified and rejected. So, the rejection probability $1-\beta$ is called the confidence level of sampling scheme at $p=p_1$. It's obvious that α and β should be minimized, generally $\alpha=0.05$ and $\beta=0.1$, and β can be 0.05 in case of higher requirements.
The values of p_0 and p_1 should be determined by the producer and consumer, and the p_1 value is generally determined by the consumer according to the

maximum lot tolerance percent defective. The values of p_0 and p_1 would be better to meet $p_1 \geq 3p_0$.

1.2.3 DETERMINATION METHOD OF SAMPLING INSPECTION SCHEME

It's described below in case of single sampling inspection by attributes.

(1) Sampling Inspection Scheme Determined by Parameters p_0, p_1, α and β

The following relations can be listed according to Figure 3-5:

$$L(p_0)=1-\alpha \tag{3-4}$$
$$L(p_1)=\beta \tag{3-5}$$

In case of single sampling inspection by attributes, when the total number of product $N>10n$, the Equations (3-4) and (3-5) can be written as

$$\sum_{r=0}^{A_C} p(r,n \mid p_0) = \sum_{r=0}^{A_C} C_n^r p_0^r q_0^{n-r} = 1-\alpha \tag{3-6}$$

$$\sum_{r=0}^{A_C} p(r,n \mid p_1) = \sum_{r=0}^{A_C} C_n^r p_1^r q_1^{n-1} = \beta \tag{3-7}$$

It can be seen from Equations (3-6) and (3-7) that there are one-to-one correspondence between the parameters p_0, p_1, α, β and the single sampling inspection by attributes. For users' convenience in application, the statisticians have calculated some tables of single sampling inspection by attributes. So the user can directly find the n and A_c values of the corresponding single sampling inspection by attributes in the tables simply according to the determined parameters p_0, p_1, α and β. Tables 3-1 and 3-2 are most commonly used for single sampling inspection by attributes.

(2) Sampling Inspection Scheme Determined According to Parameters p_1 and β

The sampling inspection scheme is determined simply according to the Equation (3-7). It's obvious that the unknown n and A_c cannot be solved by only one equation, so one of the parameters should be first determined by other principles, and then the other parameter can be solved by the Equation (3-7).

It's obvious that determining the sampling inspection method according to the p_1 and β values (also called LTPD method) emphasizes the consumer's interest, but does not take account of the producer's risk.

1.3 RELIABILITY SAMPLING OF ELECTRICAL APPARATUS FOR EXPONENTIAL DISTRIBUTION

The requirement for electrical apparatus reliability mainly concerns the examinations of product failure rate, mean life or reliable life. The reliability sampling inspection to examine failure rate is called failure rate sampling inspection. The reliability sampling inspection to examine mean life is called mean life sampling inspection. The reliability sampling inspection to examine reliable life is called reliable life sampling inspection.

1.3.1 FAILURE RATE SAMPLING

(1) OC-curve of Failure Rate Sampling Inspection Scheme

To the electrical apparatus with reliability index, the reliability is mostly represented by failure rate, and the reliability level is appraised by sampling inspection, now the OC-curve of sampling scheme represents the relation between the acceptable probability and the actual product failure rate. For discrimination, R (Reliability) is added before the OC-curve in some documents, i.e. it's called ROC-curve. The ideal ROC-curve should be step, but the kind of ideal ROC-curve cannot be realized, and the actual ROC-curve is shown in Figure 3-6.

In the figure, λ_0 is called acceptable failure rate, generally abbreviated as AFR, in some documents λ_0 is also called acceptable reliability level, abbreviated as ARL. When the actual product failure rate has $\lambda \leq \lambda_0$, the batch of products should be considered acceptable. The rejection probability is α at $\lambda=\lambda_0$ owing to misjudgment of unqualified. α is called producer's risk.

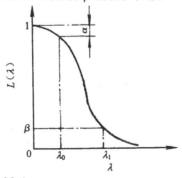

Figure 3-6 ROC-curve of failure

Table 3-1 Table of single sampling inspection by attributes

Table 3-1 Table of single sampling inspection by attributes ($\alpha=0.05$, $\beta=0.1$)

$p_0(\%)$ \ $p_1(\%)$	0.71–0.90	0.90–1.12	1.13–1.40	1.41–1.80	1.81–2.24	2.25–2.80	2.81–3.55	3.56–4.50	4.51–5.60	5.61–7.10	7.11–9.00	9.01–11.2	11.3–14.0	14.1–18.0	18.1–22.4	22.5–28.0	28.1–35.5	$p_1(\%)$
0.090–0.112	•	400 1	300 1	300 1	250 1	60 0	60 0	50 0	50 0	40 0	30 0	30 0	25 0	20 0	15 0	15 0	10 0	0.090–0.112
0.113–0.140	•	500 2	300 1	250 1	250 1	200 1	50 0	50 0	40 0	40 0	30 0	25 0	25 0	20 0	15 0	15 0	10 0	0.113–0.140
0.141–0.180	•	500 2	400 2	250 1	200 1	200 1	150 1	40 0	40 0	30 0	30 0	25 0	20 0	20 0	15 0	15 0	10 0	0.141–0.180
0.181–0.224	•	•	400 2	300 2	200 1	150 1	150 1	120 1	30 0	30 0	25 0	25 0	20 0	15 0	15 0	10 0	10 0	0.181–0.224
0.225–0.280	•	•	500 3	300 2	150 1	150 1	120 1	120 1	100 1	24 0	20 0	20 0	15 0	15 0	15 0	10 0	10 0	0.225–0.280
0.281–0.355	•	•	•	400 3	200 2	150 1	120 1	100 1	100 1	80 1	20 0	15 0	15 0	15 0	10 0	10 0	10 0	0.281–0.355
0.356–0.450	•	•	•	500 4	200 2	150 2	120 2	100 1	80 1	80 1	60 1	15 0	15 0	15 0	10 0	7 0	7 0	0.356–0.450
0.451–0.560	•	•	•	•	300 3	150 2	120 2	100 2	80 1	60 1	60 1	50 1	16 0	10 0	10 0	7 0	7 0	0.451–0.560
0.561–0.710	•	•	•	•	400 4	250 3	120 2	100 2	90 2	60 1	50 1	50 1	40 1	10 0	7 0	7 0	5 0	0.561–0.710
0.711–0.900	•	•	•	•	500 6	300 4	150 3	100 2	100 2	90 2	60 2	40 1	40 1	30 2	25 1	7 0	5 0	0.711–0.900
0.901–1.12	•	500 3	•	•	•	300 6	200 4	120 3	150 4	90 3	60 2	50 2	40 1	25 1	25 1	20 1	5 0	0.901–1.12
1.13–1.40	300 1	•	•	•	600 10	600 10	250 6	150 4	100 3	90 3	50 2	50 2	30 1	25 1	20 1	20 1	15 1	1.13–1.40
1.41–1.80	•	•	•	•	•	•	400 10	250 6	200 6	120 4	80 3	50 2	40 2	25 1	20 1	15 1	15 1	1.41–1.80
1.81–2.34	•	•	•	•	•	•	•	•	300 10	150 6	100 4	60 3	50 3	30 2	25 2	15 1	15 1	1.81–2.34
2.25–2.80	•	•	•	•	•	•	•	•	•	250 10	120 6	70 4	50 3	40 3	25 2	20 2	10 1	2.25–2.80
2.81–3.55	•	•	•	•	•	•	•	•	•	•	200 10	100 6	60 4	40 3	25 2	20 2	15 2	2.81–3.55
3.56–4.50	•	•	•	•	•	•	•	•	•	•	•	150 10	80 6	50 4	30 3	20 2	15 2	3.56–4.50
4.51–5.60	•	•	•	•	•	•	•	•	•	•	•	•	120 10	60 6	40 4	25 3	15 2	4.51–5.60
5.61–7.10	•	•	•	•	•	•	•	•	•	•	•	•	•	100 10	50 6	30 4	20 3	5.61–7.10
7.11–9.00	•	•	•	•	•	•	•	•	•	•	•	•	•	•	70 10	40 6	25 4	7.11–9.00
9.01–11.2	•	•	•	•	•	•	•	•	•	•	•	•	•	•	•	60 10	30 6	9.01–11.2
$p_1(\%)$ \ $p_0(\%)$	0.71–0.90	0.90–1.12	1.13–1.40	1.41–1.80	1.81–2.34	2.25–2.80	2.81–3.55	3.56–4.50	4.51–5.60	5.61–7.10	7.11–9.00	9.01–11.2	11.3–14.0	14.1–18.0	18.1–22.4	22.5–28.0	28.1–35.5	

Note: 1. The items marked • in the table can be referenced to Table 3-2.

2. In every blank filled with number, the number on the left represents the n value of the corresponding single sampling inspection scheme, and the number on the right represents the A_c value.

Table 3-2 - Supplementary table of single sampling inspection by attributes

$\dfrac{P_1}{P_0}$	A_c	n
1.86~1.99	20	$\dfrac{7.04}{P_0}+\dfrac{13.50}{P_1}$
2.0~2.2	15	$\dfrac{5.02}{P_0}+\dfrac{10.65}{P_1}$
2.3~2.7	10	$\dfrac{3.08}{P_0}+\dfrac{7.70}{P_1}$
2.8~3.5	6	$\dfrac{1.64}{P_0}+\dfrac{5.27}{P_1}$
3.6~4.3	4	$\dfrac{0.985}{P_0}+\dfrac{4.00}{P_1}$
4.4~5.5	3	$\dfrac{0.683}{P_0}+\dfrac{3.34}{P_1}$
5.6~7.8	2	$\dfrac{0.409}{P_0}+\dfrac{2.66}{P_1}$
7.9~16	1	$\dfrac{0.178}{P_1}+\dfrac{1.94}{P_1}$

In figure 3-6, λ_1 is called lot tolerance failure rate, generally abbreviated as LTFR; when the actual product failure rate $\lambda>\lambda_1$, the batch of products should be considered rejected, the acceptable probability β is called consumer's risk owing to misjudgment of acceptable when $\lambda=\lambda_1$.

(2) Determination Method of Failure Rate Sampling Inspection Scheme

Parameter α is generally valued 0.05 or 0.1, β is valued 0.1, λ_0 is determined according to the ability of the producer and the quality requirement of the user and other factors, λ_1/λ_0 is generally valued 1.5~5. We need to determine a scheme according to the given $\lambda_0,\lambda_1,\alpha,\beta$ for failure rate sampling inspection. Since the failure rate sampling inspection mostly adopts the single sampling inspection by attributes, the method to determine failure rate sampling scheme is to determine the number of samples n and allowable failure operations (acceptance number) A_c according to the given the values of λ_0, λ_1, α and β.

When the life of product is subjected to exponential distribution of single parameter, we can obtain approximately

$$1 - \sum_{r=0}^{A_C} C_n^r (\lambda_0 t_g)^r (1 - \lambda_0 t_g)^{n-r} = \alpha \tag{3-8}$$

$$\sum_{r=0}^{A_C} C_n^r (\lambda_1 t_g)^r (1 - \lambda_1 t_g)^{n-r} = \beta \tag{3-9}$$

where t_g is the test time. When $n\lambda t_g < 5$ and $\lambda t_g < 0.1$, the Equations (3-8) and (3-9) can be expressed by the following

$$1 - \sum_{r=0}^{A_C} \frac{e^{-n\lambda_0 t_g} (n\lambda_0 t_g)^r}{r!} = \alpha \tag{3-10}$$

$$\sum_{r=0}^{A_C} \frac{e^{-n\lambda_1 t_g} (n\lambda_1 t_g)^r}{r!} = \beta \tag{3-11}$$

Parameter n and A_c of the failure rate sampling scheme can be determined by the given λ_0, λ_1, α and β.

However in failure rate sampling inspection, the so-called λ_1 or LTFR scheme is usually adopted, i.e. the sampling scheme is determined only by Equation (3-11) according to the given λ_1 and β.

1.3.2   MEAN LIFE SAMPLING

In mean life sampling inspection, the ROC-curve of sampling inspection represents the relation between the acceptable probability and the actual mean life of product, as shown in Figure 3-7. θ_0 is called acceptable mean life level, and θ_1 is called lot tolerance mean life.

In mean life sampling inspection, the sampling scheme is usually determined by the given θ_1 and β. When the life of product is subjected to exponential distribution of single parameter, failure rate of the product is the reciprocal of its mean life, so the following relation between λ_1 and θ_1 is:

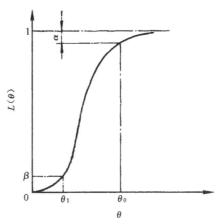

Figure 3-7 ROC-curve of mean life sampling inspection scheme

$$\lambda_1 = \frac{1}{\theta_1} \tag{3-12}$$

Simply putting the given θ_1 into Equation (3-12), we can obtain the corresponding λ_1, and then determine the sampling scheme by Equation (3-11).

1.3.3 RELIABLE LIFE SAMPLING

If the required reliable life of electrical apparatus is t_{Rg}, the maximum failure rate λ_{max} of electrical apparatus can be obtained by the following equation:

$$\lambda_{max} = \frac{1}{t_{Rg}}(-\ln R) \tag{3-13}$$

We can determine the reliable life sampling inspection scheme by putting the λ_{max}, the given confidence level and test time t_g into Equation (3-11).

1.4 RELIABILITY SAMPLING OF ELECTRICAL APPARATUS FOR WEIBULL DISTRIBUTION

The reliability sampling of products in case of Weibull distribution mainly includes mean life sampling and reliable life sampling. The so-called mean life sampling refers to the sampling inspection to examine whether the mean life of product meets the specified requirement, and the reliable life sampling refers to the sampling inspection to examine whether the reliable life of product meets the specified requirement.

The test method of reliability sampling commonly adopts timed truncated test without replacement (test deadline t_c) and single sampling inspection scheme, and the acceptable probability is a function of $F(t)$, so it can be expressed by $L[F(t)]$,

$$L[F(t)] = \sum_{r=0}^{A_c} C_n^r F(t)^r [1 - F(t)]^{n-r} \tag{3-14}$$

Equation (3-14) is the basic relation equation to determine the product reliability sampling scheme, which mostly adopts the method corresponding to the LTPD method to determine the sampling scheme.

In many cases, the position parameter v of Weibull distribution is zero, and the mean life sampling and reliable life sampling are discussed below in case of $v = 0$ only.

1.4.1 MEAN LIFE SAMPLING

Mean life sampling is to determine the sample size n, acceptance number A_c and test deadline t_c of single sampling scheme according to the specified mean life value. In case of $v = 0$, the acceptable probability is :

$$L(\mu) = \sum_{r=0}^{A_c} C_n^r \left\{ 1 - e^{-[\frac{t_c}{\mu}\Gamma(1+\frac{1}{m})]^m} \right\}^r \left\{ e^{-[\frac{t_c}{\mu}\Gamma(1+\frac{1}{m})]^m} \right\}^{n-r} \tag{3-15}$$

The relation curve between $L(\mu)$ and μ is shown in Figure 3-8.

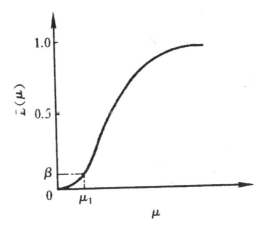

Figure 3-8 Relation curve between $L(\mu)$ and μ in mean life sampling scheme

In Figure 3-8, μ_1 is lot tolerance mean life, i.e. the required minimum mean life value; β is user's risk. $1-\beta$ is usually called the confidence level of sampling scheme. It can be seen from Figure 3-8 that $L(\mu_1) = \beta$ when $\mu = \mu_1$, so we can obtain

$$\sum_{r=0}^{A_c} C_n^r \left\{ 1 - e^{-[\frac{t_c}{\mu_1}\Gamma(1+\frac{1}{m})]^m} \right\}^r \left\{ e^{-[\frac{t_c}{\mu_1}\Gamma(1+\frac{1}{m})]^m} \right\}^{n-r} = \beta \qquad (3\text{-}16)$$

The sample size n should be selected by taking into consideration the batch size N, product price, test period, test equipment capacity and other factors etc., and it is generally determined by referencing the Chinese National Standard GB/T2828.1. The larger the batch size N and the larger sample size n, the higher the price. After n is determined, if t_c is selected, A_c can be determined by Equation (3-16), so the sampling scheme is accordingly determined; on the contrary. If A_c , the acceptance number is selected, then t_c can be determined by Equation (3-16), and the sampling scheme is accordingly determined. A_c should not be too small because the smaller A_c, the smaller the acceptable probability of sampling scheme when the actual mean life of product is specified.

By Equation (3-16), the mean life sampling table can be calculated when the confidence level $1-\beta$ and the shape parameter m of Weibull distribution both are different values. When confidence $1-\beta = 0.9$, and shape parameter $m = 0.5$, 1.0, 1.5 and 2.0, the mean life sampling table is shown in Table 3-3 to 3-6.

Table 3-3 Mean life sampling table for $m=0.5$

t_s/μ_s A_c \ n	0	1	2	3	4	5
2	0.6627	4.4097				
3	0.2945	1.3295	5.6666			
5	0.1060	0.3844	0.9799	2.3918	7.4897	
8	0.0414	0.1359	0.2985	0.5674	1.0203	1.8940
13	0.0157	0.0486	0.0994	0.1726	0.2746	0.4158
20	0.0066	0.0199	0.0394	0.0658	0.1000	0.1436
32	0.0026	0.0076	0.0148	0.0241	0.0357	0.0498
50	0.0011	0.0031	0.0059	0.0095	0.0139	0.0191
80	0.00041	0.0021	0.0023	0.0036	0.0053	0.0072

Table 3-4 Mean life sampling table for $m=1$

t_s/μ_s A_c \ n	0	1	2	3	4	5
2	1.1513	2.9697				
3	0.7675	1.6307	3.3665			
5	0.4605	0.8768	1.3998	2.1872	3.8703	
8	0.2878	0.5213	0.7727	1.0653	1.4286	1.9183
13	0.1151	0.3118	0.4459	0.5875	0.7411	0.9119
20	0.1771	0.1996	0.2807	0.3627	0.4473	0.5360
32	0.0720	0.1235	0.1718	0.2194	0.2671	0.3156
50	0.0461	0.0786	0.1087	0.1378	0.1667	0.1956
80	0.0288	0.0489	0.0674	0.0851	0.1025	0.1197

Table 3-5 Mean life sampling table for $m=1.5$

t_s/μ_s A_c \ n	0	1	2	3	4	5
2	1.2168	2.2887				
3	0.9286	1.5347	2.4882			
5	0.6606	1.0148	1.3862	1.8665	2.7307	
8	0.4829	0.7175	0.9327	1.1554	1.4051	1.7102
13	0.3494	0.5093	0.6466	0.7770	0.9072	1.0417
20	0.2622	0.3784	0.4749	0.5634	0.6479	0.7309
32	0.1916	0.2747	0.3424	0.4029	0.4595	0.5135
50	0.1423	0.2032	0.2522	0.2956	0.3355	0.3732
80	0.1040	0.1482	0.1834	0.2144	0.2426	0.2691

Table 3-6 Mean life sampling table for $m=2$

A_c / t_c/μ_1 n	0	1	2	3	4	5
2	1.2107	1.9445				
3	0.9886	1.4409	2.0703			
5	0.7657	1.0566	1.3350	1.6688	2.2199	
0	0.6054	0.0147	0.9919	1.1646	1.3486	1.5628
13	0.4749	0.6300	0.7535	0.8649	0.9714	1.0775
20	0.3829	0.5042	0.5979	0.6795	0.7547	0.8261
32	0.3027	0.3966	0.4677	0.5285	0.5832	0.6339
50	0.2421	0.3163	0.3719	0.4189	0.4607	0.4990
80	0.1914	0.2496	0.2929	0.3292	0.3613	0.3905

1.4.2   RELIABLE LIFE SAMPLING

Reliable life sampling is to determine the sample size n, acceptance number A_c and test deadline t_c of single sampling scheme according to the given reliable life value.

In case of $\nu=0$, we can obtain

$$L(t_R) = \sum_{r=0}^{A_c} C_n^r \left\{ 1 - e^{(\frac{t_c}{t_R})^m \ln R} \right\}^r \left\{ e^{(\frac{t_c}{t_R})^m \ln R} \right\}^{n-r} \tag{3-17}$$

The relation curve between $L(t_R)$ and t_R is shown in Figure 3-9.

In Figure 3-9, t_{R1} is lot tolerance reliable life. Similar to mean life sampling, $1-\beta$ is called the confidence level of reliable life sampling.

It can be seen from Figure 3-9 that $L(t_{R1}) = \beta$ at $t_R = t_{R1}$. Equation (3-18) can be obtained by putting it into (3-17)

$$\sum_{r=0}^{A_c} C_n^r \left\{ 1 - e^{(\frac{t_c}{t_{R1}})^m \ln R} \right\}^r \left\{ e^{(\frac{t_c}{t_{R1}})^m \ln R} \right\}^{n-r} = \beta \tag{3-18}$$

In reliable life sampling, the determination method of sample size is the same to that in mean life sampling. After n is determined, A_c can be determined by Equation (3-18) according to the given t_c, and the sampling scheme is determined accordingly; on the contrary, if A_c is selected, the test deadline t_c can be determined by Equation (3-18). When confidence level $1-\beta = 0.9$, and shape parameter m respectively equals to 0.5, 1.0, 1.5 and 3.0, the reliable life sampling table at reliability of 0.9 is shown in Tables 3-7 to 3-10.

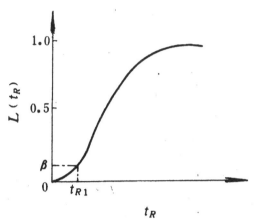

Figure 3-9 Relation curve between $L(t_R)$ and t_R of reliable life sampling scheme

Table 3-7 Reliable life sampling table for $m=0.5$

t_c/t_{R1} A_c n	0	1	2	3	4	5
2	119.40	794.48				
3	53.080	239.54	1020.9			
5	19.104	69.255	176.52	430.93	1349.4	
8	7.4627	24.479	53.780	102.23	183.83	331.50
13	2.8261	8.7550	17.914	31.089	49.481	74.909
20	1.1940	3.5898	7.0992	11.849	18.025	25.876
32	0.4664	1.3744	2.6597	4.3358	6.4290	8.9742
50	0.1910	0.5594	1.0635	1.7114	2.5034	3.4448
80	0.0746	0.2157	0.4090	0.6528	0.9468	1.2916

Table 3-8 Reliable life sampling table for $m=1$

t_c/t_{R1} A_c n	0	1	2	3	4	5
2	10.9272	28.1846				
3	7.2848	15.4770	31.9521			
5	4.3709	8.3220	13.2862	20.7588	36.7341	
8	2.7318	4.9477	7.3335	10.1110	13.5585	18.2071
13	1.6811	2.9589	4.2325	5.5757	7.0343	8.6550
20	1.0927	1.8947	2.6644	3.4422	4.2456	5.0869
32	0.6829	1.1723	1.6308	2.0822	2.5355	2.9957
50	0.4371	0.7459	1.0313	1.3082	1.5822	1.8560
80	0.2732	0.4644	0.6395	0.8079	0.9731	1.1365

Table 3-9 Reliable life sampling table for $m=1.5$

t_c/t_{R1} A_c n	0	1	2	3	4	5
2	4.9242	9.2618				
3	3.7579	6.2105	10.0693			
5	2.6733	4.1066	5.6096	7.5533	11.0504	
8	1.9542	2.9036	3.7746	4.6759	5.5860	6.9209
13	1.4138	2.0010	2.6166	3.1444	3.6713	4.2155
20	1.0609	1.5312	1.9219	2.2798	2.6220	2.9578
32	0.7755	1.1118	1.3855	1.6306	1.8594	2.0761
50	0.5759	0.8225	1.0207	1.1961	1.3578	1.5103
80	0.4210	0.5997	0.7423	0.8675	0.9820	1.0890

Table 3-10 Reliable life sampling table for $m=3$

t_c/t_{R1} A_c n	0	1	2	3	4	5
2	2.2131	3.0433				
3	1.9385	2.4921	3.1732			
5	1.6350	2.0265	2.3685	2.7483	3.3242	
8	1.3979	1.7040	1.9428	2.1624	2.3845	2.6308
13	1.1890	1.4356	1.6176	1.7732	1.9161	2.0532
20	1.0300	1.2374	1.3863	1.5099	1.6193	1.7198
32	0.8806	1.0544	1.1771	1.2770	1.3636	1.4416
50	0.7589	0.9069	1.0103	1.0937	1.1653	1.2289
80	0.6489	0.7744	0.8616	0.9314	0.9909	1.0436

2 RELIABILITY DESIGN THEORY OF ELECTRICAL APPARATUS

The reliability design is a design method to take the product reliability into consideration, its purpose is to achieve higher product reliability under specified conditions (such as cost, weight, volume and energy consumption, etc.), or to make lower cost (less weight, volume and energy consumption) under the condition of ensuring a certain reliability level by the reliability engineering method.

As stated above, the product reliability is classified into inherent reliability and operational reliability. The inherent reliability is mainly ensured by the product reliability design during the design stage, so the product reliability design is an important link which producer provides the quality guarantee for users, and the product reliability level depends on the level of product reliability design of a great extent.

Generally there are two methods of product reliability design as follows:
(1) Design with Unspecified Product Reliability Index
In this case, we can design several schemes by the conventional design method, then predict the reliability of each design scheme, and select the optimal one.
(2) Design with Predetermined Product Reliability Index
In this case, we first distribute the product reliability indexes to the parts or components or the used electronic elements of the product, then conduct reliability technical design (during which, the requirement of product performance and cost need to be considered), and conduct reliability prediction at last. If the predicted value of the reliability characteristic parameter (such as reliability, failure rate or mean life, etc.) does not meet the predetermined product reliability index, we need to improve the product reliability. In order to make the reliability characteristic of product meet the specified requirement, we need to find and improve the weak points of system reliability by reliability analysis generally.

The first design method mentioned above is generally suitable for the product with insufficient reliability data accumulated, while the second design method is used for the product with sufficient reliability data accumulated, by which the product reliability index can be predetermined.

It can be seen from the two above methods of reliability design that the content of product reliability design mainly includes the product reliability technical design, product reliability prediction, product reliability distribution and product reliability analysis, as well as the reliability design of main parts or components of electrical apparatus.

2.1 RELIABILITY PREDICTION

Reliability prediction refers to measuring the reliability characteristic parameters (such as reliability, failure rate or mean life, etc.) according to the reliability data of the parts or components of the products.
The objectives of reliability prediction as follows:
(1) The designer can estimate the reliability of the new designed product as early as possible, and determine whether it achieves the predetermined index: if not, the measures can be taken to improve as early as possible; if the predetermined

index exceeds too much, the measures can be taken to regulate the reliability level of parts or components to reduce cost and weight of the products timely.

(2) To select the optimal scheme by comparing the reliability level of every design scheme.

During product reliability prediction, the reliability data of the parts or components (or the used electrical elements) of the products must be known, which are mainly statistically obtained and accumulated in practice, or inquired from relevant documents and manuals.

A product is usually composed of many parts or components. For example, an electromagnetic relay is composed of the electromagnetic system and contact system, both of which are composed of several parts respectively. For another example, an electronic time relay is composed of several electronic elements and a small electromagnetic relay. So during product reliability prediction, the product can be viewed as a system, while the parts or components of the product can be viewed as units (elements or subsystems) that compose this system.

The reliability of a product (system) is related to the reliability level of its parts or components (units) on one hand, on the other hand it is related to the influence of the working state of these parts or components (units) on the working state of the whole product. So when the reliability of a product (system) is predicted, the reliability model of the product (system) is usually established firstly. The system reliability model means the relation between the system reliability and the reliability of its each unit. This relation can be expressed by the logic block diagram of product (system) reliability (called reliability block diagram for short), which is used for the reliability prediction.

2.1.1 RELIABILITY BLOCK DIAGRAM OF SYSTEM

The reliability block diagram refers to the block diagram that expresses the dependency of all composing parts of the product to complete a task successfully. The reliability block diagram differs from the general circuit diagram or schematic. It expresses the functional relationship among the units in the system. The reliability block diagram is drawn on the basis of the influence of normal work or failure of the units on the system working state. Each block represents a unit (element or subsystem)in the diagram. When all of the units from left to right are working normally on every pathway, the system is in normal working state; otherwise the system is in failure state.

The reliability block diagram of system also differs from the function block diagram of system in the following respects:

(1) The reliability block diagram only express the logical relation of each unit in terms of reliability, does not express the physical and time relation between the units. Therefore, the arrangement of units is not strictly ordered as the function block diagram. For example, an electromagnetic relay is composed of electromagnetic system and contact system, so it's obvious that the electromagnetic relay can work normally only in case of normal work of these two subsystems. We can draw the reliability block diagram of electromagnetic relay which is shown in the Figure 3-10.

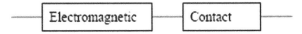

Figure 3-10 Reliability block diagram of electromagnetic relay

(2) In some cases, the units of a system are physically parallel and accordingly they are parallel in the function block diagram, but series in the reliability block diagram. For example, the 3-phase contacts of a contactor are physically parallel, but the relations between the contacts are series in the reliability block diagram.

(3) The multiple functional requirements of the same system are difficult to be expressed in the function block diagram respectively, but the reliability logical relation of each unit with different functional requirements must be expressed in the reliability block diagram.

A system can be classified into series systems, parallel systems, k-out-of-n system, mixed configurations, and complex systems, etc. according to the influence of the working state of its composing units on the system working state.

2.1.2 RELIABILITY PREDICTION OF SERIES SYSTEMS

(1) Definition and Reliability Block Diagram of Series systems
If a system can work normally only in case that all units are normal in the system (in other words, any failure of unit in the system will lead system failure), the system is called series systems, the reliability block diagram is shown as Figure 3-11. Most electrical apparatus can be viewed as series systems.

Figure 3-11 Reliability block diagram of series system

(2) Reliability Prediction of Series systems
Suppose the reliability of the ith unit is $R_i(t)(i=1,2,..., n)$, the reliability of n units in series is then

$$R_s(t) = \prod_{i=1}^{n} R_i(t)$$ (3- 19)

When the life of each unit obeys exponential distribution and the failure rate of the ith unit is noted as λ_i, the failure rate of the series system λ_s is then

$$R_s(t) = \prod_{i=1}^{n} e^{-\lambda_i t} = \exp\left[-\sum_{i=1}^{n} \lambda_i t\right] = e^{-\lambda_s t}$$ (3- 20)

$$\lambda_s = \sum_{i=1}^{n} \lambda_i$$ (3- 21)

The mean life of series system is

$$MTTF_s = \frac{1}{\lambda_s} = \frac{1}{\sum_{i=1}^{n} \lambda_i}$$ (3- 22)

If the failure rate of each unit is equal, i.e. $\lambda_1 = \lambda_2 =...= \lambda_n = \lambda$, then

$$\lambda_s = n\lambda$$ (3- 23)

$$MTTF_s = \frac{1}{n\lambda}$$ (3- 24)

2.1.3 RELIABILITY PREDICTION OF PARALLEL SYSTEMS

(1) Definition and Reliability Block Diagram of Parallel Systems
If the system can work normally in case one unit in the system works normally (in other words, the system failure results from the situation that all of the units are faulted), the system is called parallel systems, the reliability block diagram is shown as Figure 3-12.

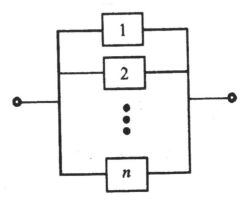

Figure 3-12 Reliability block diagram of parallel system

(2) Reliability Prediction of Parallel Systems
The reliability of n units in parallel can be calculated by

$$R_s(t) = 1 - \prod_{i=1}^{n} \left[1 - R_i(t)\right]$$

(3- 25)

If $R_1(t)=R_2(t)=\cdots=R_n(t)=R(t)$, then $R_s(t)=1-[1-R(t)]^n$.

If the life of each unit in system obeys exponential distribution, i.e. $R_i(t)=e^{-\lambda_i t}$, then

$$R_s(t) = 1 - \prod_{i=1}^{n} (1 - e^{-\lambda_i t})$$

(3- 26)

The mean life of parallel system is

$$MTTF_s = \int_0^{\infty} R_s(t)dt = \int_0^{\infty} \left[1 - \prod_{i=1}^{n} (1 - R_i(t)) \right] dt$$

(3- 27)

$$MTTF_S = \sum_{i=1}^{n} \frac{1}{\lambda_i} - \sum_{1 \le i < j \le n} \frac{1}{\lambda_i + \lambda_j} + \cdots + (-1)^{n-1} \frac{1}{\lambda_1 + \lambda_2 + \cdots + \lambda_n}$$

(3- 28)

If the failure rate of each unit is equal, i.e. $\lambda_1 = \lambda_2 = \ldots = \lambda_n = \lambda$

$$MTTF_s = \frac{1}{\lambda} \left[C_n^1 - \frac{C_n^2}{2} + \frac{C_n^3}{3} - \cdots + (-1)^{n-1} \frac{C_n^n}{n} \right]$$

(3- 29)

If a parallel system is composed of two equal units with exponential reliability (i.e. $R_1(t)=R_2(t)=e^{-\lambda t}$), the system's reliability $R_s(t)$, density function $f_s(t)$, failure rate $\lambda_s(t)$ and $MTTF_s$ are given respectively by

$$
\begin{cases}
R_s(t) = 1 - \prod_{i=1}^{n} (1 - e^{-\lambda_i t}) = 1 - (1 - e^{-\lambda t})^2 = e^{-\lambda t}(2 - e^{-\lambda t}) \\[2mm]
f_s(t) = -\dfrac{dR_s(t)}{dt} = 2\lambda e^{-\lambda t} - 2\lambda e^{-2\lambda t} = 2\lambda e^{-\lambda t}(1 - e^{-\lambda t}) \\[2mm]
\lambda_s(t) = \dfrac{f_s(t)}{R_s(t)} = \dfrac{2\lambda e^{-\lambda t}(1 - e^{-\lambda t})}{e^{-\lambda t}(2 - e^{-\lambda t})} = \dfrac{2\lambda(1 - e^{-\lambda t})}{2 - e^{-\lambda t}} \\[2mm]
MTTF_s = \dfrac{3}{2\lambda} = \dfrac{3}{2}\theta
\end{cases}
\tag{3-30}
$$

When the life of each unit obeys exponential distribution, the failure rate $\lambda_s(t)$ of the system is not a constant, but a function of operating time t.

2.1.4   RELIABILITY PREDICTION OF *K*-OUT-OF-*N* SYSTEMS

(1)   Definition and Reliability Block Diagram of *k*-out-of-*n* systems
If the system can work normally when *k* units work normally in the system of *n* units, it is called *k*-out-of-*n* system, and its reliability block diagram is shown as Figure 3-13.

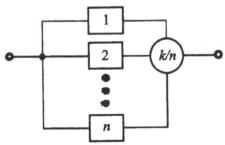

Figure 3-13 Reliability block diagram of *k*-out-of-*n* system

For example, a motor can be firmly fixed when only three of four anchor bolts work normally, so the system of anchor bolts of the motor is a *k*-out-of-*n* system with *n* = 4 and *k* = 3.

(2)   Reliability Prediction of *k*-out-of-*n* systems

To a k-out-of-n system, the normal work probability of the system (i.e. system reliability) should be equal to the sum of the following probabilities: normal work probability of n units, normal work probability of $n-1$ units, $\cdots$, and normal work probability of k units. If the normal work probability of all units (i.e. reliability) is equally $R(t)$, the system reliability is

$$R_s(t) = C_n^n R(t)^n + C_n^{n-1} R(t)^{n-1}[1 - R(t)] + C_n^{n-2} R(t)^{n-2}[1 - R(t)]^2$$
$$+ \cdots + C_n^k R(t)^k [1 - R(t)]^{n-k}$$

$$= \sum_{i=k}^{n} C_n^i R(t)^i [1 - R(t)]^{n-i} \tag{3-31}$$

If the life of each unit is subject to exponential distribution, i.e. $R(t) = e^{-\lambda t}$, wherein λ is the failure rate of each unit, the system reliability is

$$R_s(t) = \sum_{i=k}^{n} C_n^i (e^{-\lambda t})^i [1 - e^{-\lambda t}]^{n-i} = \sum_{i=k}^{n} C_n^i e^{-i\lambda t}[1 - e^{-\lambda t}]^{n-i} \tag{3-32}$$

$$MTTF_s = \int_0^{\infty} [\sum_{i=k}^{n} C_n^i e^{-i\lambda t}(1 - e^{-\lambda t})^{n-i}]dt \tag{3-33}$$

It can be proved by induction that $C_n^i \int_0^{\infty} [e^{-i\lambda t}(1 - e^{-\lambda t})^{n-i}]dt = \dfrac{1}{i\lambda}$, and by

putting this relation into the Formula (3-33), we can obtain

$$MTTF_s = \sum_{i=k}^{n} \frac{1}{i\lambda} = \frac{1}{k\lambda} + \frac{1}{(k+1)\lambda} + \cdots + \frac{1}{n\lambda} \tag{3-34}$$

2.1.5   RELIABILITY PREDICTION OF SERIES-PARALLEL SYSTEMS

(1) Definition and Reliability Block Diagram of Series-Parallel Systems
The system with both series and parallel connections is called series-parallel system. Figure 3-14 is the reliability block diagram of a series-parallel system which is composed of 5 units.

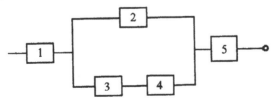

Figure 3-14 Reliability block diagram of series-parallel system

(2) Reliability Prediction of Series-Parallel Systems

1) System Reliability Calculated by Simplified Method

The electrical apparatus with some parts of the circuit using redundancy design is a series-parallel system. Every series-parallel system can be viewed as a system composed of some subsystems (subsystems in series or parallel connections). When predicting the reliability of series-parallel systems, obtain the reliability of subsystems by the prediction method introduced above, i.e. and simplify the reliability block diagram of series-parallel systems to a simple system (series systems or parallel systems) gradually, and obtain the reliability of the whole system.

2) System Reliability Range Estimated by the Cut and Tie Sets

The so-called cut sets is to draw a line through some blocks in the reliability block diagram of the system, and the failure of these blocks will lead to system failure; the so-called tie sets is to draw a line through some blocks in the reliability block diagram of system, and the normal work of these blocks will lead to normal work of the system.

The range of system reliability R_s can be obtained by the following formula:

$$1 - \sum_{i=1}^{N} \prod_{j=1}^{k_i} (1 - R_j) < R_S < \sum_{i=1}^{T} \prod_{j=1}^{m_i} R_j \qquad (3\text{-}35)$$

where N = the number of cut sets;

k_i = the unit number on the ith split line;

T = the number of tie sets;

m_i = the unit number on the ith connected line

The more complex the systems, the more complex are the calculations of reliability R_s of the system. An approximate value (range) of R_s can be more rapidly obtained by the above algorithm of cut and tie sets.

2.1.6   RELIABILITY PREDICTION OF COMPLEX SYSTEMS

(1) Definition and Reliability Block Diagram of Complex Systems

In practical problem, there is a type of complex network system which is neither series nor parallel, and difficult to be simplified to simple series-parallel system, referred to as complex system.

Such as the reliability block diagram of a complex system shown as Figure 3-15, it's called bridge system because its diagram is similar to a bridge.

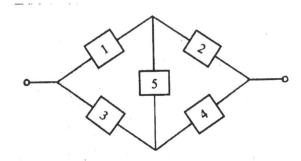

Figure 3-15 Reliability block diagram of bridge system

(2) Reliability Prediction of Complex systems

1) Truth Table Method (Contingency Enumeration Method)
Suppose the system is composed of n units, "1" represents normal work state of each unit and the system, m represents the number of system state value "1"; "0" represents failure state, and l represents the number of system state value "0". The number of system states combination N is

$$N = m + l = 2^n$$

The system reliability R_s is the sum of the probabilities of system state values "1", i.e.

$$R_s = \sum_{i=1}^{m} P(S_i = 1) \qquad (3\text{-}36)$$

2) Application of Boolean Algebra Algorithm
Basic Relations of Boolean Algebra: Boolean function refers to the relation $f(x_1, x_2, x_3,...)$ obtained from union ($\cup$), intersection ($\cap$), negation ($\bar{\ }$) and other operations performed by the Boolean variables $x_1, x_2, x_3, ...$.
Expansion Theorem: Suppose $y = f(x_1, x_2, ..., x_n)$, the above Boolean function is f_1 when $x_i = 1$ and the above Boolean function is f_0 when $x_i = 0$. To any Boolean variable x_i, the Boolean function y can be expanded as

$$y = f_1 x_i + f_0 \bar{x}_i \qquad (3\text{-}37)$$

Bridge System Reliability Calculated by Expansion Theorem:
If the units 1, 2, 3, 4, 5 in Figure 3-15 are expressed by Boolean variables x_1, x_2, x_3, x_4, x_5 respectively, the Boolean function of the bridge system is

$$y = f(x_1, x_2, x_3, x_4, x_5) \qquad (3\text{-}38)$$

By expanding the Formula (3-38) for x_5, we obtain

$$y = x_5 f_1 + \bar{x}_5 f_0 \qquad (3\text{-}39)$$

where f_1 = the Boolean function in the Formula (3-38) when $x_5 = 1$, which can also be expressed as $f(x_1, x_2, x_3, x_4, 1)$

f_0 = the Boolean function in the Formula (3-38) when $x_5 = 0$, which can also be expressed as $(x_1, x_2, x_3, x_4, 0)$.

So the Formula (3-39) can be written as

$$y = x_5 f(x_1, x_2, x_3, x_4, 1) + \bar{x}_5 f(x_1, x_2, x_3, x_4, 0) \qquad (3\text{-}40)$$

where $(x_1, x_2, x_3, x_4, 1)$ represents the Boolean function with x_5 always being 1(i.e. the unit 5 will not fail), so it can be expressed by the reliability block diagram shown as Figure 3-16.

It can be obtain from logical relation:

$$f(x_1, x_2, x_3, x_4, 1) = (x_1 + x_3)(x_2 + x_4) \qquad (3\text{-}41)$$

Similarly, $(x_1, x_2, x_3, x_4, 0)$ represents the Boolean function with x_5 always being 0 (i.e. the unit 5 is always in failure), so it can be expressed by the reliability block diagram shown as Figure 3-17.

It can be obtained from logical relation:

$$f(x_1, x_2, x_3, x_4, 0) = x_1 x_2 + x_3 x_4 \qquad (3\text{-}42)$$

By putting the Formulas (3-41) and (3-42) into the Formula (3-40), we obtain

$$y = x_5 (x_1 + x_3)(x_2 + x_4) + \bar{x}_5 (x_1 x_2 + x_3 x_4) \qquad (3\text{-}43)$$

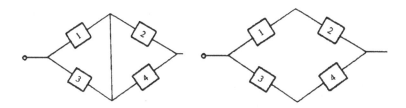

Figure 3-16 Reliability block diagram of bridge system with normal unit 5

Figure 3-17 Reliability block diagram of bridge system with failed unit 5

It's obvious that the bridge system reliability R_s is the probability of the Boolean function y which is equal to 1, if the units 1, 2, 3, 4, 5 normal are represented by events A_1, A_2, A_3, A_4, A_5 respectively, and the system normal is represented by event A_s, the Formula (3-43) can also be expressed by the relation between events.

$$A_s = A_5(A_1 + A_3)(A_2 + A_4) + \overline{A_5}(A_1A_2 + A_3A_4) \qquad (3\text{-}44)$$

It's obvious that the events $A_5(A_1 + A_3)(A_2 + A_4)$ and $\overline{A_5}(A_1A_2 + A_3A_4)$ in the formula are mutually exclusive. Then take

$$P(A_5) = R_5$$

$$P(A_1 + A_3) = P(A_1) + P(A_3) - P(A_1A_3) = R_1 + R_3 - R_1R_3$$

$$P(A_2 + A_4) = P(A_2) + P(A_4) - P(A_2A_4) = R_2 + R_4 - R_2R_4$$

$$P(\overline{A_5}) = 1 - P(A_5) = 1 - R_5$$

$$P(A_1A_2 + A_3A_4) = P(A_1A_2) + P(A_3A_4) - P(A_1A_2A_3A_4) = R_1R_2 + R_3R_4 - R_1R_2R_3R_4$$

which can be put into the Formula (3-44) to obtain the bridge system reliability:

$$R_s = R_5(R_1 + R_3 - R_1R_3)(R_2 + R_4 - R_2R_4) + (1 - R_5)(R_1R_2 + R_3R_4 - R_1R_2R_3R_4)$$
$$(3\text{-}45)$$

Put $R_1 = 0.8$, $R_2 = 0.7$, $R_3 = 0.8$, $R_4 = 0.7$, $R_5 = 0.9$ into the Formula (3-45) to obtain $R_s = 0.86688$. Comparing to the true table method, the advantage of this method is that we can not only obtain the value of system reliability, but also obtain the relational expression of system's reliability.

2.2 RELIABILITY DISTRIBUTION

Product (system) reliability distribution is to make the reliability indexes of the units (elements or subsystems) according to the predetermined product (system) reliability indexes, i.e. to distribute the predetermined product (system) reliability indexes to every units of product. Reliability distribution is conducted generally at the design and reliability improvement of the product with reliability indexes requirement. Its purpose is to specify the reliability requirement of the composing units of product (system), and to realize this requirement by taking measures at reliability technical design.

System reliability distribution is to distribute the reliability indexes of system design requirements to the subsystems, devices and elements in the system by specified method, and write them into the corresponding design requirements as the design basis. Reliability distribution is a decomposition process from whole to part, from larger to smaller, and from top to bottom ones.

The purpose of reliability distribution is to make the design personnel be familiar with the requirements of reliability design, accordingly estimate the required manpower, time and resources, etc, and study the possibility and methods to realize the reliability indexes. Reliability distribution is mainly applied in the stage of scheme demonstration and engineering development.

In order to improve the justification and feasibility of reliability distribution, follow the rules below during reliability distribution:

(1) System or product with high complexity is distributed with lower reliability index;

(2) Product with mature technology and good succession is distributed with higher reliability index;

(3) Product applied in severe environment is distributed with lower reliability index;

(4) Product with long task time is distributed with lower reliability index;

(5) Product with high importance is distributed with higher reliability index;

(6) Reliability distribution is a process of decision-making, other constraints should be considered to realize optimal system design.

Several common methods of reliability distribution are introduced below.

2.2.1   SIMPLE METHOD OF RELIABILITY DISTRIBUTION (EQUAL DISTRIBUTION METHOD)

Equal distribution method is based on the principle of equal reliability (R) of each unit of the product. This distribution method is simple for calculation and easy for application, suitable in the stage of scheme verification and scheme design. Its main disadvantage is that the actual differences of each subsystem are not taken into consideration.

(1)   Reliability Distribution of Series Systems

Suppose the predetermined product (system) reliability is R_s, the reliability distributed to every unit is R. In the series systems composed of n units, we have

$$R = \sqrt[n]{R_s} \tag{3-46}$$

(2)   Reliability Distribution of Parallel Systems

To the parallel systems composed of n units, we have

$$R = 1 - \sqrt[n]{1 - R_s} \tag{3-47}$$

(3)   Reliability Distribution of Series-Parallel Systems

Taking the parallel parts in the system as one unit, and distribute the reliability by the way of distribution method for series systems; then distribute the reliability to parallel parts by the way of equal distribution method; and so on, finally obtain the distributed value of every unit.

2.2.2 RELIABILITY DISTRIBUTION ACCORDING TO RELATIVE FAILURE RATE

This method distributes the system reliability to every composing unit according to their relative failure. It is suitable for the new system because it is very similar to the original system, and the reliability of the original system and its composing units are known or can be inferred. For a series system of n units, if the failure rate of each unit is estimated as $\hat{\lambda}_1, \hat{\lambda}_2, \ldots, \hat{\lambda}_n$ according to the existing data, and the system failure rate is inferred as

$$\hat{\lambda}_s = \sum_{i=1}^{n} \hat{\lambda}_i \tag{3-48}$$

Then the relative failure rate of the ith unit is

$$\omega_i = \frac{\hat{\lambda}_i}{\hat{\lambda}_s} = \frac{\hat{\lambda}_i}{\sum\limits_{i=1}^{n} \hat{\lambda}_i} \tag{3-49}$$

If the predetermined product (system) failure rate index is λ_s, the failure rate distributed to the ith unit is

$$\lambda_i = \omega_i \lambda_s \tag{3-50}$$

Suppose the predetermined reliability of the product (system) that working to the given moment t_g is $R_s(t_g)$, the reliability distributed to every composing unit is

$$R_i(t_g) = [R_s(t_g)]^{\omega_i} \tag{3-51}$$

2.2.3 RELIABILITY DISTRIBUTION METHOD WITH COST MINIMIZATION

This method is suitable for the conditions of optimal design. The so-called distribution method with minimum cost is to realize minimum overall development cost under ensuring overall product reliability index. It can solve the most important and practical problems in reliability design.

To apply this method, first establish the relation between reliability and development cost of the unit

$$R_i = f(C_i) \tag{3-52}$$

And then establish the relation between system reliability and unit reliability

$$R = \prod_{i=1}^{n} R_i \tag{3-53}$$

So the problem of reliability distribution comes down to: obtaining the unit reliability R_i which would minimize $C = \sum_{i=1}^{n} C_i$ and satisfy the system reliability R.

2.3 RELIABILITY TECHNICAL DESIGN

2.3.1 DERATING USE

The so-called derating use refers to the "plan to reduce the internal stress of material or element to improve the reliability". To the electronic components, it is generally referred to the reduction of the power consumed (or voltage applied) in actual application of the components comparing to the rated power (or rated voltage) of the element by a certain amplitude.

It can be known from Arrhenius equation that the element life will increase with the decrease of temperature T. Therefore, when the relation between the electronic element life and temperature complies with Arrhenius equation, and the power consumed in actual application is lower than the rated power, its temperature is also lower than the temperature at rated power. As a result, its life will be longer than the life at rated power, consequently, its failure rate is reduced and reliability is improved. Likely, when the relation between electronic element life and the applied voltage complies with inverse power law equation, and the applied voltage in actual application is lower than the rated voltage, its life will also be prolonged, consequently, its failure rate is reduced and reliability is improved.

When performing the product design, the actual derating degree of the adopted electronic element depends on the actual situation. For example, in case of small number of used elements, and it's not very difficult to achieve the specified product reliability, we should only ensure it will not exceed the rated value in application, without the necessity for derating with significant amplitude; in case of large number of used elements, and it's difficult to achieve the specified product reliability, we should conduct derating by significant amplitude.

2.3.2   RESERVE DESIGN (REDUNDANCY DESIGN)

The so-called reserve design (redundancy design) method is to reserve some units with the same function as standbys, and to improve the whole system or equipment reliability. The typical reserve method is parallel method; for example, in case a product is composed of n units, it can work normally if one of its units can work normally, so this reserve method is parallel.

It should be noticed that the definition of the above-mentioned parallel method differs from that of parallel circuit. For example, if the main failure mode of a certain capacitor is short-circuit, the original one capacitor can be replaced by two same capacitors in serial connection to improve its reliability, as shown in Figure 3-18(a). In this case, short-circuit will not happen if only one of the two capacitors has no short-circuit; short-circuit will happen only when both capacitors have short-circuit. If the probability of one capacitor short-circuit is represented by P_1, the probability of both capacitor short-circuit is P_1^2. It's obvious that $P_1^2 < P_1$, so replacing the original one capacitor by two capacitors in series can reduce the probability of short-circuit and to improve the circuit reliability, which is a reserve design method adopting parallel method (although the two capacitors are in series connection).

If the main failure mode of the capacitor is open-circuit, the reserve design method of two parallel capacitors should be adopted as shown in Figure 3-18(b). Now open-circuit will happen only when both capacitors have open-circuit, so the probability of open-circuit is reduced, and the circuit reliability is improved. If the capacitor comes with failure modes of both short-circuit and open-circuit, the circuit shown in Figures 3-18(a) and (b) cannot be used to improve the circuit reliability, while hybrid reserve circuit can be adopted as shown in Figure 3-18(c). Now open-circuit will happen only when both capacitors C_1 and C_2 or both C_3 and C_4 have open-circuit; short-circuit will happen when one of C_1 and C_2 and one of C_3 and C_4 have short-circuit; so the probability of open-circuit and short-circuit is reduced, and the circuit reliability is improved.

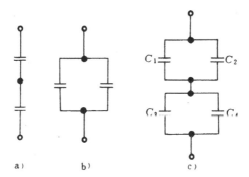

Figure 3-18 Reserve design of capacitor circuit (a) Two capacitors in series (b) Two capacitors in parallel (c) Four capacitors with two-two in parallel and then in series

2.3.3 ENVIRONMENT RESISTANCE DESIGN

Environment resistance design is the design taking account of various environmental conditions, including mechanical stress (impact and vibration, etc.) resistance design, climatic conditions (high temperature, low temperature, humidity and salt mist) resistance design, and radiation resistance design, etc. In environment resistance design, the problem to be considered is the environment resistance measures in the design and the environmental conditions for the expected actual application of the product.

2.3.4 HEAT RESISTANCE DESIGN

The design to control the temperature rise at each part of product is called heat resistance design. We can take measures to improve the heat resistance performance and to improve the product heat dispersion by strengthening convection and radiation. The core problem of heat resistance design is thermodynamic calculation.

2.3.5 VIBRATION RESISTANCE DESIGN

(1) Influence of Vibration and Impact
Impact is generated when applying sudden acting force on an object, with possibly significant velocity. In general cases, impact will cause temporary object vibration with frequency equal to the object's inherent vibration frequency

(which depends on the object's weight and structure), and amplitude is related to the friction and other resistances. The smaller the friction, the larger is the amplitude of the vibration frequency. Impact may also lead to the following consequences:

1) Damage to fragile material;

2) Deformation of soft material; and

3) Malfunction of product.

Vibration is the force that is periodically applied with alternative different strength. It may lead to product fatigue failure. Resonance will be generated when the vibration frequency equals to the inherent vibration frequency of product part, which may lead to complete product damage.

(2)   Vibration resistance measures

Common vibration resistance measures:

Appropriate element configuration (or installation): installation of heavier element on the lower part of the overall product; firm fixation of electrical wiring on fixed base plate or framework, which shall not be loose near the leading-out end; adopting spring washer for bolt fixation; do not use outrigger as far as possible.

Adoption of buffering: in order to reduce the vibration and impact transmitted from the base (or framework and base plate, etc.) of the product, appropriate vibration absorber can be installed between the base and product. For impact resistance, the buffering material (such as rubber, etc.) with elasticity under compression, or air buffer can be used. For vibration resistance, cellular paper vibration absorber, polystyrene foam plastic blocks, rubber mat, and metal spring, etc. can be used to reduce the transmission of vibrating energy, and to reduce the product inherent vibration frequency lower than the lower limit of the externally applied vibration source to avoid resonance.

2.4 RELIABILITY DESIGN OF ELECTRICAL CONTACT

Contact is the key part that influences the life and reliability of electrical apparatus. According to statistics, over 70% of small relay faults are caused by contact fault. So the contact reliability is extremely important for electrical apparatus, which is not only closely related to the application environment, contact circuit parameters, etc, but also determined by the reliability design to a great extent. The development of reliability design of electrical contact is of extreme significance.

2.4.1 SELECTION OF MATERIAL AND STRUCTURE FORM OF ELECTRICAL CONTACT

(1) Selection of Structure Form of Contact

The contact resistance is composed of constriction (or pinch) resistance and film resistance. The constriction resistance is formed by the pinching of the line of current near the contact, while the film resistance is formed by various membranes of dust film, absorbed film, organic membrane and inorganic membrane, etc. on the contact surface. The surface membrane of large and medium load contact can be crushed or burned out by the electric arc or spark generated when the contact current is disconnected, so its contact resistance is mainly determined by the pinch resistance; therefore it's better to choose surface or line contact with more contacts. The surface membrane of small load contact cannot be easily crushed, so its contact resistance is mainly determined by the film resistance. In order to improve the capacity for cleaning the surface membrane by increasing the contact pressure, it's better to choose point contact with fewer contacts.

(2) Selection of Contact Material

The resistivity, hardness and chemical property of contact material can directly influence the value of contact resistance. So the large load should be made by the material with high hardness, high melting point, good electric and thermal conductivity, and high ultimate arc parameter, while the small load should be made by the material with less contact resistance and good chemical stability. In addition, it should be taken account of selecting appropriate organic material and adopting appropriate technique to minimize the gas hazardous to contact generated in application, and adopting the material that is difficult to generate oxide film, organic membrane and absorbed film.

The electrical contacts made with different materials are for matching the switching applications. This can limit metal transfer caused by liquid bridge, which is especially important for the contact of small relays. The directional metal transfer caused by liquid bridge is one of the causes for electric erosion of contacts. In order to limit the metal transfer caused by liquid bridge, the positive pole can be made by the material with good electric and thermal conductivity, and the negative pole can be made by the material with poor electric and thermal conductivity, as shown in Table 3-11.

Table 3-11 Matching table of contact material

Material of positive pole	Ag	Ag	Ag	Ag	Au	Au	Au
Material of negative pole	Au	Pt	Pd	Pt-Ir	Pt	Pd	Pt-Ir

Matching of different materials will also improve the electric abrasion resistance of contact. For example, when Ag-CdO and Ag are matched, the electric abrasion rate is lower than that of both contacts made by pure Ag-CdO, and it is much lower than that of contact pair made by pure Ag. Another example when Ag-Ni (containing Ni20%) and Ag are matched, the electric abrasion rate is also lower than that of both contacts made by pure Ag-Ni or by pure Ag.

2.4.2 DESIGN METHOD FOR REDUCING AND STABILIZING CONTACT RESISTANCE

(1)   Appropriate increase of contact pressure: when the contact pressure is increased, the constriction (or pinch) resistance and film resistance will reduce, so appropriate increase of contact pressure is an effective method to reduce and stabilize the contact resistance. However, the increase of contact pressure should not be excessive, we should consider the factors of matching characteristics and the size of electromagnetic system, etc.

(2)   Correct selection of contact surface processing precision: to the milli-ampere level small load contact, in order to reduce pollution, the methods such as polishing, can be adopted to achieve lower surface roughness of the contact.

(3)   The contact used in the circuit with low voltage and current should be designed to come with the structure of certain relative sliding and rolling as much as possible.

(4)   In order to eliminate the high contact resistance between plug and socket, do not use plug connection as much as possible.

(5)   Prevention of contact pollution: attention should be paid as much as possible to the prevention of contact pollution during assembly, transport, storage and installation of electrical apparatus; the electrical apparatus should be

occupied with dust hood; the electrical contact should be separated from the organic insulating material as far as possible; the contact applied in the environment with erosive chemical gas should adopt sealed structure; the getter materials that can absorb the organic gas volatilized from the insulating material can be put in the sealed relay.

(6)　In order to prevent contact fault caused by entry of dust between the contacts, the double- terminal contact as shown in Figure 3-19 can be adopted. It's shown by test result that the fault probability can be significantly reduced by adopting double-terminal contact at the same total contact pressure.

Figure 3-19 Double-terminal Contact

2.4.3　DESIGN METHOD TO REDUCE BOUNCE

Contact bounce will cause increase of contact electric wearing, and even lead to contact splicing, consequently influence the electrical apparatus reliability. The common method to reduce contact bounce is as follows:

(1) Appropriately increasing the contact initial pressure;

(2) Appropriately reducing the contact closing velocity;

(3) Appropriately reducing the mass of movable contact;

(4)　To make the moving directions of contact and armature perpendicular to each other as much as possible in case of product design;

(5)　To avoid contact bounce caused by armature shaking when the electrical apparatus is impacted by external force, we can adopt a completely balanced structure of armature, reduce the mass of armature and improve the ratio between the force applied on the armature and the armature weight;

(6) Adopting rotating structure and changing the lever ratio;

(7)　Adopting the collision type iron core. It can reduce the colliding velocity of iron core and produce buffering to reduce the contact bounce caused by iron core impact.

2.4.4   REDUNDANCY DESIGN OF CONTACT CIRCUIT

In the situation with higher requirement for contact operational reliability, redundancy electrical contact design can be adopted to improve the contact circuit operational reliability.

When the main fault mode of contact is excessive contact resistance (poor contact), the circuit of redundancy design is with parallel contacts as shown in Figure 3-20. The resistance between terminal a and b is excessive only in case of poor contact of both contacts K_1 and K_2. In case which the poor contact probability of one contact is P_k, the poor contact probability of both contacts is P_k^2, and it's obvious $P_k^2 < P_k$. So when the main fault mode of contact is poor contact, the circuit of redundancy design as shown in Figure 3-20 can be adopted to improve the contact operational reliability.

When the main fault mode of contact is short-circuit (i.e. splicing), the circuit of redundancy design is as shown in Figure 3-21.   In this case, short-circuit between terminal a and b occurs only in case of short-circuit of both contacts. If the short-circuit probability of one contact is P_d, the short-circuit probability of both contacts is P_d^2. It's obvious $P_d^2 < P_d$. So when the main fault mode of contact is short-circuit, redundancy design as shown in Figure 3-21 can be adopted to improve the contact operational reliability.

When the contact has both fault modes of poor contact and splicing (the probabilities of the two are equal or approximate), neither circuits as shown in Figures 3-20 and 3-21 is capable for improving the contact operational reliability. The circuit of redundancy design as shown in Figure 3-22 can be adopted, i.e. 4 contacts in parallel two by two and then in series to be used as one.

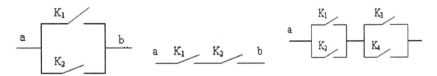

Figure 3-20
Parallel circuit of contacts

Figure 3-21
Series circuit of contacts

Figure 3-22
Four contacts with two-two in parallel and then in series

The resistance between ab is excessive only in case of poor contacts at both K_1 and K_2 or both K_3 and K_4; short-circuit between ab occurs only in case of splicing at least at one of K_1 and K_2 and one of K_3 and K_4. So the probability of poor contact or short-circuit between ab can be reduced and the contact operational reliability is improved.

The fault probability and reliability of the redundancy design circuit as shown in Figure 3-22 are discussed below. Suppose the failure distribution of contact is exponential, with failure rate of λ, the reliability $R(t)$ and failure rate $F(t)$ of one contact are respectively

$$R(t) = e^{-\lambda t} \tag{3-54}$$

$$F(t) = 1 - e^{-\lambda t} \tag{3-55}$$

Suppose the ratio between the probability of poor contact and the probability of short-circuit (splicing) keeps constant, and the ratio of poor contact probability to the contact total fault probability is k, the poor contact probability $F_k(t)$ and short-circuit probability $F_d(t)$ of one contact are respectively

$$F_k(t) = k(1 - e^{-\lambda t}) \tag{3-56}$$

$$F_d(t) = (1 - k)(1 - e^{-\lambda t}) \tag{3-57}$$

For different fault modes of contact, the reliability block diagrams of the redundancy design circuit shown in Figure 3-22 are different. When the contact fault mode is poor contact, its reliability block diagram is shown as Figure 3-23.

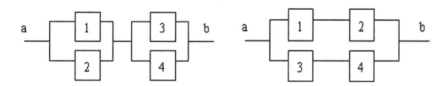

Figure 3-23
Reliability block diagram of contact fault mode of poor contact

Figure 3-24
Reliability block diagram of contact failure mode of short-circuit

So the fault probability of poor contact between terminals a and b is

$$F_{sk}(t) = 1 - [1 - F_k^2(t)]^2 \tag{3-58}$$

When the contact fault mode is short-circuit, its reliability block diagram is shown as Figure 3-24. So the fault probability of short-circuit between ab is

$$F_{sd}(t) = \{1 - [1 - F_d(t)]^2\}^2 \tag{3-59}$$

The total fault probability and reliability of the circuit with redundancy design shown in Figure 3-22 are respectively

$$F_s(t) = F_{sk}(t) + F_{sd}(t) = 1 - [1 - F_k^2(t)]^2 + \{1 - [1 - F_d(t)]^2\}^2 \tag{3-60}$$

$$R_s(t) = 1 - F_s(t) = [1 - F_k^2(t)]^2 - \{1 - [1 - F_d(t)]^2\}^2 \tag{3-61}$$

By putting Formulae (3-56) and (3-57) into the Formula (3-61), we obtain

$$R_s(t) = [1 - k^2(1 - e^{-\lambda t})^2]^2 - \{1 - [1 - (1 - k)(1 - e^{-\lambda t})]^2\}^2 \tag{3-62}$$

When knowing the value of k and contact failure rate λ, we can obtain the reliability at any work moment of the circuit with redundancy design as shown in Figure 3-22 by the Formula (3-62).

2.5 RELIABILITY DESIGN OF MECHANICAL PART IN ELECTRICAL APPARATUS

2.5.1 BASIC PRINCIPLE OF RELIABILITY DESIGN BASED ON LOAD-STRENGTH INTERFERENCE MODEL

The load-strength interference model, abbreviated as interference model, can be used to clearly disclose the nature of part reliability design, and it's the basic model for part reliability design.

In general, the physical quantities applied on the product or part, such as stress, pressure, temperature, humidity, impact and voltage, etc. are called the load endured by the product or part, denoted by Y; the degrees of the product or part to endure this stress are generally called the strength of the product or part, denoted by X. If the strength of product or part X is less than the load Y, the product or part will fail to complete the specified function, which is called failure. The product or part is capable of reliable work within a specified time only when

$$Z = X - Y \geq 0 \tag{3-63}$$

In fact, the strength X and load Y are functions of some variables, i.e.

$$X = f_X(X_1, X_2, \cdots, X_m)$$
$$Y = g_Y(Y_1, Y_2, \cdots, Y_n) \tag{3-64}$$

where, X_i Is the stochastic variable that influences strength, such as the material

properties, surface quality and size effect of the part; Y_i is the stochastic

quantity that influences load, such as the loading condition, stress concentration, working temperature and lubricating state, etc.

So both load and strength are stochastic variables subjected to certain distribution. Since the load and strength have the same dimensions, they can be expressed in the same coordinate system. The load-strength interference model is shown as Figure 3-25.

It can be known from the nature of statistic distribution function that the two probability density function curves of load and strength may intersect under certain conditions, which is also called interference, and this intersection phenomenon is called interference phenomenon. Their overlap area (as Figure 3-25) is called interference area in this model. As to mechanic parts, even there is no interference phenomenon in the early stage, the part strength will be reduced with time or with cyclic loading. For example, when the part strength varies from $t=0$ to $t=t_0$ along the degradation curve, interference will be occur between the probability density function curves of the load and strength in Figure 3-25. In the interference area, the load maybe exceeds the strength and failure will occur.

At certain moment t, its load-strength interference is shown as Figure 3-26, in which, the abscissa indicates the load and the strength. The curves $f(x)$

and $g(y)$ indicate the strength and load probability densities respectively. The shadow in the figure is the "interference area" of the load-strength densities distribution, indicating the strength may become less than load in this area. This model calculating product reliability according to the interference situation of the load and strength is called the load-strength interference model.

It can be seen from the interference diagram that:

(1)   In case of higher dispersion degree of part strength and working load, the interference area must be enlarged, and the unreliability is increased;

(2)   In case of good material performance and stable load, the dispersion degree of both density functions will be decreased, the interference area will be reduced, and the reliability will be increased. It indicates that an adequate safety factor is not sufficient in the normal approach to design and reliability is needed to be calculated.

It should be noticed that whether the part fails or not should be further analyzed in the interference area in Figure 3-26. Namely, the part fails if the load exceeds the strength. Otherwise, the part does not fail.

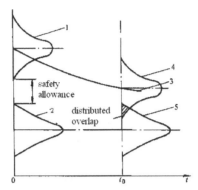

Figure 3-25 Load-strength interference model: 1—Strength distribution at $t = 0$; 2—Load distribution at $t = 0$; 3—Degradation curve of average of the strength; 4—Strength distribution at $t = t_0$; 5—Load distribution at $t = t_0$

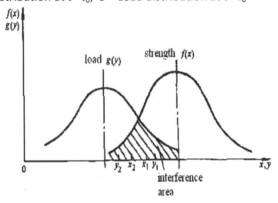

Figure 3-26 Load-strength interference diagram

If Y and X are independent stochastic variables, their difference, $Z = X - Y$, is also a stochastic variable. And the product or part reliability R is the probability when Z is more than or equal to zero, i.e.

$$R = P(Z \geq 0) \qquad (3\text{-}65)$$

Cumulative failure probability is

$$P_F = 1 - R = P(Z < 0) \qquad (3\text{-}66)$$

It can be seen from the interference model that every design has failure probability, i.e. reliability $R \leqslant 1$. So what we can do is to control the failure probability within an acceptable limit at design.

The calculation of reliability based on the model stated above generally adopts analytic method, numerical integration, graphics, Monte Carlo simulation, etc.

2.5.2   GENERAL EXPRESSION OF FAILURE PROBABILITY AND RELIABILITY CALCULATIONS

Suppose the probability density functions of strength X and Load Y respectively are $f(x)$ and $g(y)$, the cumulative distribution functions are $F(x)$ and $G(y)$, then there are two methods to determine the failure probability P_F and reliability R.

(1) Failure Probability or Reliability Calculated by Multiplication Rule for Probability

Now enlarge the interference area in Figure 3-26, as shown in Figure 3-27, and take a small interval dy from the interference area, the probability of load y in dy is

$$P(y_1 - \frac{dy}{2} \leq y \leq y_1 + \frac{dy}{2}) = g(y_1)dy \tag{3-67}$$

and the probability of strength x less than stress y_1 is

$$P(x < y_1) = \int_{-\infty}^{y_1} f(x)dx \tag{3-68}$$

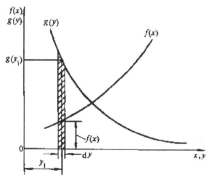

Figure 3-27   Enlarged diagram of interference area

Since both of them are random events independent of each other, it can be known by the multiplication rule of probability that the probability of simultaneous occurrence of both events equals to the product of the probabilities of independent occurrence of each of them, i.e. it equals to

$$g(y_1)dy \cdot \int_{-\infty}^{y_1} f(x)dx .$$

This probability is the interference probability in the small interval dy caused by load, i.e. failure probability. Then, regarding the load distribution as a whole, the part failure probability is

$$P_F = P(x < y) = \int_{-\infty}^{\infty} g(y)[\int_{-\infty}^{y} f(x)dx]dy \qquad (3\text{-}69)$$

Because $R = 1 - P_F$, and $\int_{-\infty}^{y} f(x)dx + \int_{y}^{\infty} f(x)dx = 1$, so the corresponding reliability is

$$R = P(x \geq y) = \int_{-\infty}^{\infty} g(y)[\int_{y}^{\infty} f(x)dx]dy \qquad (3\text{-}70)$$

P_F can also be calculated according to the probability of $y > x$. Similar to the above steps, we can obtain

$$P_F = P(y > x) = \int_{-\infty}^{\infty} f(x)[\int_{x}^{\infty} g(y)dy]dx \qquad (3\text{-}71)$$

$$R = P(y \leq x) = \int_{-\infty}^{\infty} f(x)[\int_{-\infty}^{x} g(y)dy]dx \qquad (3\text{-}72)$$

In addition, Equations (3-69) and (3-71) can also be expressed in the following forms

$$P_F = \int_{-\infty}^{\infty} F_x(y)g(y)dy \qquad (3\text{-}73)$$

and $P_F = \int_{-\infty}^{\infty} f(x)[1 - G_y(x)]dx = 1 - \int_{-\infty}^{\infty} f(x)G_y(x)dx \qquad (3\text{-}74)$

where $F_x(y)$ is the cumulative distribution function value of X at x=y; $G_y(x)$ is the cumulative distribution function value of Y at y=x.

If strength is a fixed value, suppose $x = a$, then

$$P_F = P(y > a) = \int_{a}^{\infty} g(y)dy = 1 - G_y(a) \qquad (3\text{-}75)$$

$$R = G_y(a) \qquad (3\text{-}76)$$

(2) Calculate the Failure Probability or Reliability by Joint Probability Density Function Integral

Because both strength X and load Y are stochastic variables, $Z = X - Y$ is also a stochastic variable, called the interference stochastic variable. Because X and Y are independent of each other, according to the Convolution Theorem on the joint probability density function of the difference between two random variables in probability theory, we can obtain the probability density function $h(z)$ of Z

$$h(z) = \int_{-\infty}^{\infty} f(z + y)g(y)dy \qquad (3\text{-}77)$$

In case $z \geq 0$, we should take $y \geq 0$, then $h(z) = \int_{0}^{\infty} f(z + y)g(y)dy$

$$R = \int_{0}^{\infty} h(z)dz = \int_{0}^{\infty}\int_{0}^{\infty} f(z + y)g(y)dzdy \qquad (3\text{-}78)$$

In case the load and strength maybe obey different distribution types, the specific calculation formula of reliability will be also different. The calculation formulae of reliability in case of the load and strength subjected to the same distribution type are shown in Table 3-12.

Table 3-12 Calculation formulae of reliability in case of load and strength subjected to the same distribution type

Distribution type	Load distribution	Strength distribution	Reliability
Normal distribution	$N(\mu_y, \sigma_y^2)$	$N(\mu_x, \sigma_x^2)$	$R = \Phi(z)$: $\quad z = \dfrac{\mu_x - \mu_y}{\sqrt{\sigma_x^2 + \sigma_y^2}}$
Logarithmic normal distribution	$\ln Y \sim$ $N(\mu_y, \sigma_y^2)$	$\ln X \sim$ $N(\mu_{lx}, \sigma_{lx}^2)$	$R = \Phi(z)$: $z = \dfrac{\mu_x - \mu_y}{\sqrt{\sigma_x^2 + \sigma_y^2}} \approx \dfrac{\ln \mu_x - \ln \mu_y}{\sqrt{V_x^2 + V_y^2}}$
Weibull distribution	(m_y, η_y, γ_y)	(m_x, η_x, γ_x)	$R = 1 - \int_{0}^{\infty} e^{-z} \exp\{-[\frac{\eta_x}{\eta_y} z^{1/m_x} + (\frac{\gamma_x - \gamma_y}{\eta_y})]^{m_y}\}dz$ $z = (\dfrac{x - \gamma_x}{\eta_x})^{m_x}$ Where
Exponential distribution	λ_y	λ_x	$R = \dfrac{\lambda_y}{\lambda_y + \lambda_x}$ or $\quad R = \dfrac{\mu_x}{\mu_x + \mu_y}$

In the table, except for the exponential distribution, the calculation formulae of reliability are complex. Although the cumulative standard normal distribution table can be looked up in case of normal distribution, it's still difficult for the programming in computer-aided design, so the approximate solution by numerical calculation method is usually adopted.

2.5.3 RELIABILITY OBTAINED BY NUMERICAL INTEGRATION

In case of complex load and strength distribution, it's difficult to obtain the calculation expression of reliability by the analytic method stated above. The numerical integration method is an alternative on the basic principle as follows: Supposing the strength and load cumulative distribution functions are $F(x)$ and $G(y)$ respectively as shown in Figure 3-28, and their density functions are $f(x)$ and $g(y)$, reliability R can be obtained by the Equation (3-79).

$$R = 1 - \int_{-\infty}^{\infty} f(x)[1 - G_y(x)]dx = \int_{-\infty}^{\infty} f(x)G_y(x)dx \approx \sum_{i=1}^{n-1} \Delta F_i G_i \quad (3\text{-}79)$$

Divide the abscissa in Figure 3-28 into $(n\text{-}1)$ equal portions within the effective calculation range. Calculate the value $G_i = [G(x_{i+1}) + G(x_i)]/2$ in stress distribution function in every equal interval, and the difference $\Delta F_i = F(x_{i+1}) - F(x_i)$ of strength distribution function. By putting ΔF_i and G_i into the Equation (3-79), reliability R can be calculated as

$$R = \sum_{i=1}^{n-1} \frac{1}{2}[G(x_{i+1}) + G(x_i)][F(x_{i+1}) - F(x_i)] \quad (3\text{-}80)$$

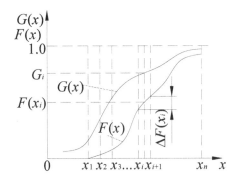

Figure 3-28   Reliability calculated by numerical integration method

2.6 RELIABILITY DESIGN FOR ELECTROMAGNETIC SYSTEM OF ELECTRICAL APPARATUS

2.6.1 BASIC PRINCIPLE OF RELIABILITY DESIGN FOR ELECTROMAGNETIC SYSTEM OF ELECTRICAL APPARATUS

The pull-in voltage is an important parameter for electrical apparatus. In order to ensure reliable work of electrical apparatus, it's specified in standards that the pull-in voltage U_x should not exceed the allowable operating voltage value U_{xy}. At design of electromagnetic system in electrical apparatus, one of constraints is that the electrical apparatus can be operated reliably when its coil is energized by a voltage. The voltage is called the designed pull-up voltage. Selecting the designed pull-up voltage should be the key at the design. If the voltage is too high, some products produced according to the designed optimal solution will be unqualified because their pull-in voltage cannot meet requirements for manufacturing error and the dispersivity of the material properties. If the voltage is too low, the mechanical life and electric life will be reduced because the contact colliding velocity and the iron colliding velocity of products will be too high when energized by rated coil voltage. Therefore, in the reliability design of electromagnetic system in electrical apparatus, the designed pull-up voltage is used as the assessment index.

In the following, the basic theory on pull-in reliability of electromagnetic system in electrical apparatus is explained, the function relation of pull-in reliability to pull-in voltage is established, and the basic method of reliability design of electromagnetic system in electrical apparatus is covered in detail.

2.6.1.1 BASIC THEORY ON PULL-IN RELIABILITY OF ELECTROMAGNETIC SYSTEM IN ELECTRICAL APPARATUS

In the production of electrical apparatus, because of the influences by some random factors (such as the machining process, tolerance and fit, magnetic material properties, and reaction force dispersivity, etc.), the pull-in voltage of the electrical apparatus has certain dispersivity, so the pull-in voltage U_x is a stochastic variable. In case it's no more than the allowable pull-in voltage U_{xy}, i.e. $U_x \leq U_{xy}$, the product is qualified. In case it's more than the allowable pull-in voltage value U_{xy}, i.e. $U_x > U_{xy}$, the product is unqualified. When the electrical apparatus reliability R is assessed by the index of pull-in voltage, the pull-in reliability R is the probability of $U_x \leq U_{xy}$, i.e.

$$R = P(U_x \leq U_{xy}) \tag{3-81}$$

The pull-in voltage should be measured according to relevant standards. After the pull-in voltage data are obtained, the distribution type can be estimated and verified according to the methods introduced in Chapter 2. Note: before the verification by K-S method, the observed values of samples should be sequenced from small to large.

In general, it can be considered the pull-in voltage is subject to normal distribution.

Suppose the pull-in voltage U_x of an electrical apparatus is subject to normal distribution with parameter (μ_{U_x} , σ_{U_x}), i.e. $U_x \sim N$ (μ_{U_x} , $\sigma_{U_x}^2$), and the allowable pull-in voltage U_{xy} is generally a multiple of the rated coil voltage U_e or a given value. According to the load-strength interference theory, the pull-in voltage U_x can be considered as load, while the allowable pull-in voltage U_{xy} can be considered as strength. The pull-in reliability $R=P(U_x \leq U_{xy})$, that is $R=P(U_x-U_{xy} \leq 0)$. It can be known from the probability theory and mathematical statistics that the variable $\xi=U_x-U_{xy}$ is also subject to normal distribution.

$$\mu_\xi = \mu_{U_x} - U_{xy} \tag{3-82}$$

$$\sigma_\xi = \sqrt{\sigma_{U_x}^2 + \sigma_{U_{xy}}^2} = \sigma_{U_x} \tag{3-83}$$

Then $z = -\dfrac{\mu_{U_x} - U_{xy}}{\sigma_{U_x}}$ (3-84)

$$R = \Phi(z) \tag{3-85}$$

where z is association coefficient. The pull-up reliability R can be got in the table of standard cumulative normal distribution (see Appendix).

According to the above basic principle, not only the pull-in voltage reliability of existing electrical apparatus can be calculated, but the electromagnetic system can also be designed by a given pull-in voltage reliability.

2.6.1.2 BASIC METHOD OF RELIABILITY DESIGN OF ELECTROMAGNETIC SYSTEM IN ELECTRICAL APPARATUS

The pull-in reliability design of Electromagnetic System in Electrical Apparatus can be divided into two class problems:

(1) First Class of Problems

For the first class of problems - calculation of the pull-in voltage reliability of existing electrical apparatus is relatively easier. We should randomly sample several units of electrical apparatus produced by different batches (considering the electrical apparatus produced by different batches may have different pull-in voltage dispersivity), measure their pull-in voltage value, and then calculate the average pull-in voltage μ_{U_x} and standard deviation σ_{U_x} by the method of mathematic statistics. The allowable pull-in voltage U_{xy} is a constant, the association coefficient z can be calculated by the Formula (3-84), and the pull-up reliability R can be got from Equation (3-85).

(2) Second Class of Problems:

The second class of problems is designing an electromagnetic system in electrical apparatus by the given pull-in reliability R. In other words, it's required that the pull-in reliability of the designed electromagnetic system should meet the requirement. This class of problems is relatively complex, and should be solved by the following three steps:

1) Determination of Variation Coefficient V_{U_x} of Pull-in Voltage

It can be known from the first class of problems that some dispersivity may exist among the pull-in voltages of the electrical apparatus produced by the same or different batches. In order to express the degree of this dispersivity, the ratio of its standard deviation to its average value is defined as the variation coefficient of pull-in voltage, denoted by V_{U_x}, i.e.

$$V_{U_x} = \frac{\sigma_{U_x}}{\mu_{U_x}} \tag{3-86}$$

where μ_{U_x} and σ_{U_x} are respectively the average value and standard deviation of pull-in voltage of electrical apparatus.

For redesign of existing products, the variation coefficient can be obtained according to the pull-in voltage data of the existing product. For design of new products, the pull-in voltage sample cannot be actually tested. It's more reasonable that the variation coefficient of pull-in voltage for the new product is determined by referencing to that of the existing products with the similar structure and production technology.

2) Calculation of Reliable Pull-in Voltage U_R

Suppose the pull-in reliability R of the electromagnetic system to be designed is a specified value, the association coefficient z can be inquired in the table of standard cumulative normal distribution (see Appendix), $\sigma_{U_x} = \mu_{U_x} \cdot V_{U_x}$ by the definition of variation coefficient, and then the average pull-in voltage μ_{U_x} is derived from Equation (3-84)

$$\mu_{U_x} = \frac{U_{xy}}{1 + z \cdot V_{U_x}} \tag{3-87}$$

where z, U_{xy} and V_{U_x} are known, so μ_{U_x} can be obtained.

The average pull-in voltage μ_{U_x} is called reliable pull-in voltage U_R because it depends on the pull-in reliability.

Taking a type of AC contactors as example, U_{xy} =323V, V_{U_x} =0.023, and if it's specified pull-in reliability R=0.99999, then z=4.265, so μ_{U_x} =294.15V, that is U_R =323V.

3) Design of Electromagnetic System

In the reliability design of electromagnetic system, U_R is used as the design value of pull-in voltage. The designed electromagnetic system is just able to operate when the coil is energized with voltage U_R, thus it can be ensured that the pull-in reliability of the designed electromagnetic system is R.

2.6.2 RELIABILITY OPTIMIZATION DESIGN OF ELECTROMAGNETIC SYSTEM IN ELECTRICAL APPARATUS

2.6.2.1 BASIC PRINCIPLE OF RELIABILITY OPTIMIZATION DESIGN OF ELECTROMAGNETIC SYSTEM IN ELECTRICAL APPARATUS

The objective functions for reliability optimization design of electromagnetic system in Electrical Apparatus can be divided into two classes of problems: one class is with the objective of the maximum pull-in reliability; the other class is with the objective of the minimum cost, volume, weight or other optimal performance indexes of the electrical apparatus under a given pull-in reliability. The constraints and solution processes of these two classes of problems are different, and are narrated respectively in detail in the following sections.

2.6.2.1.1 TAKING THE MAXIMUM PULL-IN RELIABILITY AS OBJECTIVE

(1) Objective function: the objective is maximum pull-in reliability, i.e. the major problem is to obtain R. We know $R = \Phi(z)$ is monotone increasing function, so this optimal problem can be transformed to the solution of objective function with maximum z.

It can be known from the Equation (3-84) that the higher z the lower average pull-in voltage. So this problem can be transformed to the solution of electromagnetic system with the minimum pull-in voltage, i.e. $\min \mu_{U_x}$.

(2) Constraints: constraints are the specifications that should be met by the design scheme. Different specifications should be met by the electromagnetic system with different purposes, and constrains are also different. The common constrains are introduced as follows.

1) Constraint of Coil Temperature Rise

The coil of Electromagnetic System in Electrical Apparatus consumes power and produces heat during its work, with certain temperature rise. The temperature rise exceeding a specified value will lead to coil aging, reduction of insulated performance, and even coil burning. Therefore, to ensure reliable work of electromagnetic system, the maximum temperature rise of electrical apparatus is specified in the product standard. So at design of electromagnetic system, the coil temperature rise of electrical apparatus should be less than the allowable value, i.e.

$$\tau_{max}\Big|_{\substack{\delta=\delta_{min}\\U=K_uU_e}} \le \tau_m \qquad (3\text{-}88)$$

where τ_m is the maximum allowable temperature rise; τ_{max} is the maximum temperature rise on coil; K_u is the ratio of maximum coil voltage to rated coil voltage. Because of voltage fluctuation in power grid, the temperature rise is maximum under the maximum coil voltage ($K_u U_e$), so the temperature rise should be calculated under $K_u U_e$; δ_{min} is the air gap length at closed position, i.e. the temperature rise during normal work should be calculated.

2) Constraint of Magnetic Flux Density

When the magnetic flux density exceeds a certain value, the magnetic field intensity will increase significantly because of irons with the nonlinear magnetic curve. To prevent the magnetic system from saturation state, the maximum allowable magnetic flux density B_m is specified. The magnetic system should be generally designed with the requirement that the maximum magnetic flux density $B_{max}(X)$ does not exceed the allowable value B_m, i.e.

$$B_{max}(X) \le B_m \tag{3-89}$$

$B_{max}(X)$ should be obtained through magnetic field analysis or magnetic circuit calculation under the maximum coil voltage ($K_u U_e$).

(3) Optimization Solution

The solution process of this problem is give a set of design variable values, obtain an electromagnetic system, and calculate whether this electromagnetic system meets the constraint; if not, this scheme is infeasible; we should give another set of design variable values according to optimization method, and calculate. If the constraint is met, calculate the pull-in voltage of electromagnetic system.

The pull-in voltage of electromagnetic system can calculated by single variable optimization, with the program diagram shown as Figure 3-29.

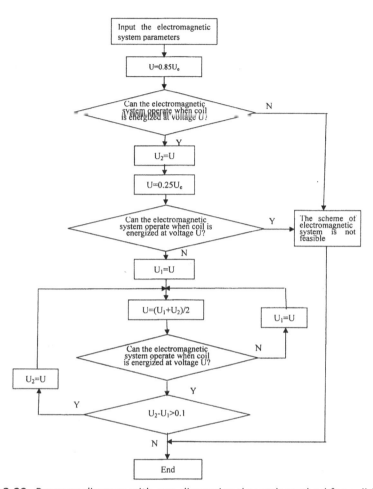

Figure 3-29 Program diagram with one-dimensional search method for pull-in voltage of electromagnetic system

If considering static case, we should judge whether the static magnetic force exceeds the reaction force. When the considering dynamic state, we should calculate the dynamic motion process of the electromagnetic system in the program, and consider whether its velocity is more than or equal to zero and other constraints.

By feasible solution of each set, a pull-in voltage value can be obtained. The design schemes may be successively given according to the optimization method, and compared, to obtain the optimal scheme, i.e. the scheme with minimum pull-in voltage and the highest pull-in reliability.

2.6.2.1.2 TAKING THE OPTIMAL PERFORMANCE AS OBJECTIVE UNDER GIVEN PULL-IN RELIABILITY

This class of problems is to design the optimal electromagnetic system to meet the requirements of pull-in reliability. We should first calculate the reliable pull-in voltage U_R, determine the objective functions and constraints, and then establish the mathematic model of reliability optimization design. The objective functions and constraints are selected as follows:

(1) Objective Functions

1) Minimum Material Volume

The material volume of electromagnetic system is the sum of iron volume $V_{Fe}(X)$ and copper conductor volume of coil $V_{Cu}(X)$, i.e.

$$V(X) = V_{Fe}(X) + V_{Cu}(X) \tag{3-90}$$

2) Minimum Weight

In the fields of traffic, transportation and aerospace, etc. it's generally required that the weight of electromagnetic system should be minimized. The following objective function can be established as designing electromagnetic system applied in these fields:

$$m(X) = m_{Fe}(X) + m_{Cu}(X) = \rho_{Fe} V_{Fe}(X) + \rho_{Cu} V_{Cu}(X) \tag{3-91}$$

where $m_{Fe}(X)$ = the ferromagnetic material weight;

$m_{Cu}(X)$ = copper conductor weight of coil;

ρ_{Fe} = ferromagnetic material density;

ρ_{Cu} = copper conductor density of coil.

3) Minimum Material Cost

The main material cost for electromagnetic system is the sum of ferromagnetic material cost and copper conductor cost of coil, here the objective function is

$$C(X) = C_{Fe}(X) + C_{Cu}(X) = \sigma_{Fe} m_{Fe}(X) + \sigma_{Cu} m_{Cu}(X) \tag{3-92}$$

where $C_{Fe}(X)$ = ferromagnetic material cost;

$C_{Cu}(X)$ = copper conductor cost of coil;

σ_{Fe} = ferromagnetic material cost per unit weight;

σ_{Cu} = copper cost per unit weight of coil.

In case the specified design requirements are met, the successful design of electromagnetic system with the minimum material cost will undoubtedly bring the manufacture direct economic benefits. Therefore, the material cost for electromagnetic system is one of the objective functions most commonly used in the optimization design of electromagnetic system.

4) Minimum Power Consumption

Electromagnetic system is the main energy consuming component in relay and contactor, and reducing the consumed power in electromagnetic system will save energy and reduce coil temperature rise. The objective function for this class of optimization design problems is

$$P(X) = P_{Fe}(X) + P_{Cu}(X) \tag{3-93}$$

where $P_{Fe}(X)$ = power consumed by iron core;

$P_{Cu}(X)$ = power consumed by coil resistance.

5) The Least Action Time of Electromagnetic System

It's generally required that the action time of electrical apparatus (particularly relay) should be minimized to meet special requirements in some application field, here the following objective function can be established:

$$t(X) = t_c(X) + t_d(X) \tag{3-94}$$

where $t_c(X)$ = actuation time of electromagnetic system;

$t_d(X)$ = motion time of electromagnetic system;

6) Minimum Kinetic Energy at Certain Position

The more kinetic energy stored in the movable iron (or armature) of magnetic system, the more colliding energy at contact or iron closing. This will cause contact or iron core bounce to reduce the mechanical life and electrical life of relay or contactor. An important measure to improve that is to ensure minimum kinetic energy for movable iron (or armature) at contact or iron impact, for which the following objective function can be established:

$$E(X) = \frac{1}{2} m(X) v^2(X) \tag{3-95}$$

where $m(X)$ = movable iron (or armature) weight;

$v(X)$ = moving velocity of movable iron (or armature) at contact or iron impact.

7) Optimal Comprehensive Index of Economic and Technical

From the viewpoint of a designer, he usually expects to obtain the electromagnetic system with all optimal economic and technical indexes. For example, it's required the electromagnetic system should come with small volume, light weight, and low material cost and consumed power. Actually. This belongs to the problem of multi-objective optimization design, which should be solved by multi-objective optimization method.

The objective functions commonly used in the optimization design of electromagnetic system are listed above. The objective functions in other forms can also be established according to the actual requirements.

(2) Constraints

Constraints are the specifications that should be met by the design scheme. Different specifications should be met by the electromagnetic system in different applications, and the constraints in the mathematic model for optimization design of electromagnetic system are also different. The objective function of a design problem may be the constraint of another design problem. However, for the optimization design problem of single objective function, the objective function and constraint in the same mathematic model of design problem should not be the same parameter. The common constraints are introduced.

1) Constraint of Static Magnetic Force

The most basic task of an electromagnetic system is to achieve the work done by driving the load, which requires the magnetic force generated by the electromagnetic system should be sufficient to overcome the reaction force so that the electromagnetic system works normal. Therefore, in traditional design, it's required that the magnetic force of electromagnetic system should be greater than the corresponding reaction force under the working air gap δ, i.e.

$$F_{x_i}(X) - K_{f_i} F_{f_i} \geq 0 \qquad i = 1, 2, \cdots, m \qquad (3\text{-}96)$$

where $F_{x_i}(X)$ = the magnetic force under the working air gap δ_i;

$F_{f_i}(X)$ = reaction force under the working air gap δ_i;

K_{f_i} = the magnetic force margin coefficient, $K_{f_i} > 1$.

In reliability optimization design, the constraint of the magnetic force is

$$F_{x_i}(X)\big|_{U=U_R} - F_{f_i} \geq 0 \qquad i = 1,2,\cdots,m \qquad (3\text{-}97)$$

2) Constraint of coil temperature rise, which is the same to Formula (3-88).

3) Constraint of magnetic flux density, which is the same to Formula (3-89).

4) Constraint of action time: it's usually required that the electromagnetic system in some electrical apparatus, especially small relays, should come with action time $T(X)$ of no more than the maximum allowable action time T_m, i.e.

$$T(X) \leq T_m \qquad (3\text{-}98)$$

5) Constraint of Colliding Velocity of Iron or Contact

The higher colliding velocity of iron or contact, the more energy generated when they collide, and consequently easier for iron or contact bounce to reduce the service life of electrical apparatus. In order to relieve iron or contact bounce, it's required the colliding velocity of the iron or contact $v(X)$ of electromagnetic system should not be more than the allowable value v_m, i.e.

$$v(X) \leq v_m \qquad (3\text{-}99)$$

6) Constraint of Iron Moving Velocity

In terms of AC electromagnetic system, the dynamic characteristics of electromagnetic system is related to the closing phase angle of coil voltage, and the characteristics of its moving velocity are different under different closing phase angles. However, it's required stagnation should not occur throughout its moving process at closing under all closing phase angles, i.e.

$$v(X)\big|_{U=U_R} \geq 0 \qquad (3\text{-}100)$$

The specifications (constraints) commonly found in general electromagnetic system design are introduced above, as to the electromagnetic systems of special type, the conditions that should be met by design variables during design process can be listed according to other specifications.

With the establishment of the mathematic model, if the selected objective function is one of the Equations （3-90） to （3-95）, it's an optimization design problem with single objective; more than one of them can also be selected, here it's an optimization design problem with multiple objectives; some constrains are forced. Therefore, the reliability optimization design of electromagnetic system in electrical apparatus is a nonlinear planning problem with constraints and single or multiple objectives.

2.6.1.2 RELIABILITY OPTIMIZATION DESIGN OF DC ELECTROMAGNETIC SYSTEM

The DC electromagnetic systems mostly come with clapper or solenoid structure, so the reliability optimization designs for these two structural forms are introduced as follows.

2.6.2.2.1 RELIABILITY OPTIMIZATION DESIGN OF DC CLAPPER ELECTROMAGNETIC SYSTEM

DC clapper electromagnetic system is the most common electromagnetic system. It is used as the sensing mechanism of most contactors and relays, with typical structure diagram as shown in Figure 3-30 and general reaction characteristic as shown in Figure 3-31. In the reliability optimization design of electromagnetic system, the original data generally given are rated coil voltage value U_e, allowable temperature rise of coil τ_m and reaction characteristic or the reaction forces at given air gaps.

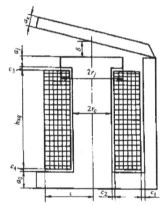

Figure 3-30 Structure diagram of clapper electromagnetic system

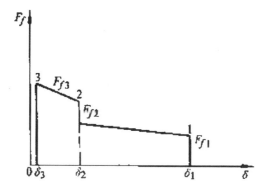

Figure 3-31 Reaction characteristic of clapper electromagnetic system

(1) Establishment of Mathematic Model

　　1) Selection of Design Variables

In the problem of engineering optimization for electromagnetic system of electrical apparatus, the key parameters are generally selected from the parameters related to geometric dimensions and coil, to be used as design variables. All other parameters can be expressed by the selected key parameters that may be: iron core radius r_c, coil height h_{xq}, coil outer radius c, and pole shoe radius r_j. So the other dimensions can be determined as follows:

Thickness of pole shoe: $a_j = \dfrac{r_c}{2}[1-(\dfrac{r_c}{r_j})^2]$

Width of magnetic yoke and armature: $b_x = b_e = 2c$

Thickness of armature: $a_x = \dfrac{(0.5 \sim 0.8)\pi r_c^2}{b_x}$

　　　In case of magnetic leakage, the magnetic flux at core end should be smaller than that in the core, so the area of iron core is multiplied by the coefficient (0.5~0.8) in the formula. The value is smaller in case of higher core and serious magnetic leakage.

Magnetic yoke thickness: $a_e = \dfrac{(1.5 \sim 2)\pi r_c^2}{b_e}$

Because of the structure or increased mechanical strength, and the largest magnetic flux at the bottom of electromagnetic system taken into consideration, its area of section is larger than the iron core.

Number of coils: $N = \dfrac{k_{tc} h_{xq} (c - r_c - c_2)}{A_d} = \dfrac{4 k_{tc} h_{xq} (c - r_c - c_2)}{\pi d^2}$

where: d = conductor diameter;

k_{tc} = fill factor of coil.

2) Establishment of Objective Function

The objective function may come with maximum pull-in reliability or other optimal index in case specified pull-in reliability is met. The latter case is introduced below. The objective may be volume, weight and cost. Both weight and cost can be calculated on the basis of iron volume and copper volume. The volumes are expressed as the function of design variables.

$$V_{Fe} = a_x b(a_e + c_1 + 2c) + \pi[r_j^2 a_j + r_c^2 (h_{xq} + c_3 + c_4)] \tag{3-101}$$
$$+ a_e b(2c + a_e + c_1 + c_4 + h_{xq} + c_3 + a_j)$$

$$V_{Cu} = \pi k_{tc}[c^2 - (r_c + c_2)^2]h_{xq} \tag{3-102}$$

The expressions of material volume, weight and cost of electromagnetic system can be obtained by Equations (3-90) to (3-92), and all of these three objective functions are the functions of design variables, excluding the reliability factor. In design, one of them can be selected as the objective function.

3) Constraints

The constraint of temperature rise and magnetic flux density are shown as Formulae (3-88) and (3-89), while the constraint of magnetic force is shown as the Formula (3-97).

The magnetic force should be calculated under reliable pull-in voltage, while both temperature rise and magnetic flux density should be calculated under maximum coil voltage. The solutions of magnetic force, temperature rise and magnetic flux density involve the calculation of electromagnetic system by magnetic circuit and field methods. The magnetic circuit method can obtain proximate formula by simplified calculation and significantly simplify the optimization problem.

In addition to performance constraints, the design variables should meet some constrains of geometrical dimensions. For example, every geometrical dimension should be more than zero, and meet

$$c + c_1 > r_j > r_c \qquad (3\text{-}103)$$

$$c > r_c \qquad (3\text{-}104)$$

4) Mathematical Model

$$
\begin{cases}
\min\{V(X), m(X), C(X)\} \\
s.t. F_{xi}\big|_{U=U_R} \geq F_{fi} \\
\quad \tau_m - \tau\big|_{U=K_u U_e} \geq 0 \\
\quad B_m - B\big|_{U=K_u U_e} \geq 0 \\
\text{Constraint of geometric dimensions}
\end{cases}
\qquad (3\text{-}105)
$$

(2) Optimization Solution

Select the appropriate optimization method to solve the mathematic model of reliability optimization design of clapper electromagnetic system.

2.6.2.2.2   RELIABILITY OPTIMIZATION DESIGN OF DC SOLENOID ELECTROMAGNETIC SYSTEM

The structure diagram of DC solenoid electromagnetic system is shown in Figure 3-32.

(1)   Selection of Design Variables

The selected design variables: armature radius r_c, coil outer radius C, arresting iron height η, coil length h_{xq}, and conductor diameter d.

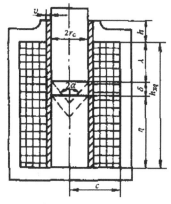

Figure 3-32 Structure diagram of DC solenoid electromagnetic system

(2) Constraints

The constraints are the same to the static reliability optimization design of DC clapper electromagnetic system, i.e. the constraint of magnetic force[the same to the Formula (3-97)], temperature rise [the same to the Formula (3-88)], and magnetic flux density [the same to the Formula (3-89)] at design points and key points. In addition to performance constraints, the design variables should meet some constrains of geometrical dimensions, for example, every geometrical dimension should be more than zero, and some mutual relations.

The magnetic force should be calculated under reliable pull-in voltage, while both temperature rise and magnetic flux density should be calculated under maximum coil voltage. The solutions of magnetic force, temperature rise and magnetic flux density of solenoid electromagnetic system also involve the calculation of the electromagnetic system.

(3) Objective Function

The effective material volume, weight and price of electromagnetic system can be used as the optimization objective. Suppose the upper and lower bottom thickness of electromagnetic system is $r_c / 2$, with equal area of section of enclosure (yoke) and core, then

$$V = \pi(c^2 + r_c^2)(h_{xq} + r_c) \tag{3-106}$$

$$m_{Cu} = k_{tc}\pi(c^2 - r_c^2)h_{xq}\rho_{Cu} \tag{3-107}$$

$$m_{Fe} = \rho_{Fe}\pi[r_c^2(2h_{xq} + r_c) + c^2 r_c] \tag{3-108}$$

$$m = m_{Cu} + m_{Fe} \tag{3-109}$$

$$c = \sigma_{Cu}m_{Cu} + \sigma_{Fe}m_{Fe} \tag{3-110}$$

The expression of mathematical model for reliability optimization design is the same to the Formula (3-105).

2.6.2.2.3 DYNAMIC RELIABILITY OPTIMIZATION DESIGN OF DC ELECTROMAGNETIC SYSTEM

If the mathematical model includes dynamic characteristic of the electromagnetic system, this kind of designs belongs to the reliability optimization design of electromagnetic system with dynamic index. In terms of the dynamic reliability optimization design of DC electromagnetic system, the selection of design variables is the same to the static reliability optimization design. The selected objective functions generally are:

Economic index of electromagnetic system——any one of volume, weight and cost, such as cost;

Colliding velocity v_1 of the dynamic contact to static contact is minimum;

Colliding velocity v_2 of the armature to iron core (or arresting iron) is minimum.

The selected constraints generally are: constraint of temperature rise, magnetic flux density, action time.

So the mathematical models are:

$$
\begin{cases}
\min C(X) \\
\min v_1(X) \\
\min v_2(X) \\
s.t. \quad g_1(X) = \tau - \tau_m \leq 0 \\
\qquad g_2(X) = B - B_m \leq 0 \\
\qquad g_3(X) = T - T_m \leq 0 \\
\text{Constraints of geometric dimensions}
\end{cases}
\tag{3-111}
$$

Of course, the objective functions and constraints may be different as particularly required.

In the mathematic model, the calculation of C, τ and B is the same to static reliability optimization design, while the other values should be calculated in dynamic process, and obtained through the calculations of the function $i(t)$, $v(t)$, $x(t)$ and $F_x(t)$ of dynamic characteristics, etc..

2.6.2.3 RELIABILITY OPTIMIZATION DESIGN OF AC ELECTROMAGNETIC SYSTEM

The iron core of AC electromagnetic system should be made by laminating silicon-steel sheets, so their structure mostly adopts one of the types U, double-U, E or double-E, etc..

The method to establish the mathematic model of its reliability optimization design is narrated as follows in the case of double-E AC electromagnetic system (as shown in Figure 3-33).

2.6.2.3.1 STATIC RELIABILITY OPTIMIZATION DESIGN OF AC ELECTROMAGNETIC SYSTEM

(1) Selection of Design Variables

According to Figure 3-33, the following parameters can be selected as the optimization design variables: $b_z, a, b_{xq}, h_{xq}, m, n, b, c, \gamma_2, d$

where $\gamma_2 =$ the ratio of the area of magnetic pole sectional in the shading coil to the total area of magnetic pole section;

$d =$ coil conductor diameter.

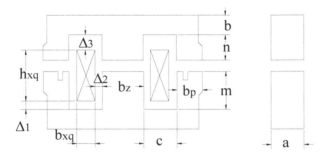

Figure 3-33 Structure diagram of double-E AC electromagnetic system

During optimization design, relative design variables are mostly adopted, to let $x_1=b_z/\delta_0, x_2=a/b_z, x_3=b_{xq}/b_z, x_4=h_{xq}/b_z, x_5=m/b_z, x_6=n/b_z, x_7=b/b_z, x_8=c/b_z, x_9=\gamma_2$, and $x_{10}=d$

where: $\delta =$ air gap length when movable core is in open position.

(2) Objective Function

For selection of optimization objective, energy and material should be saved as far as possible on the premises of reliable pull-in and release of electromagnetic system. Effect of saving energy and material is measured by the material cost, power consumed by coil in rated thermal state, volume or weight. Therefore, the objective function in optimization can be selected from these parameters.

Iron and coil volumes of electromagnetic system obtained by Figure 3-33:

$$V_{Fe} = a(b_z + 2b_p)(m+n) + 2ab(2b_p + 2c + b_z) \tag{3-112}$$

$$V_{Cu} = [2(a+b_z + 2\Delta_2)b_{xq} + \pi b_{xq}^2]h_{xq} \tag{3-113}$$

then total volume $V = V_{Cu} + V_{Fe}$ (3-114)

and total weight $m = m_{Cu} + m_{Fe} = k_{Fe}\rho_{Fe}V_{Fe} + \rho_{Cu}V_{Cu}$ (3-115) where: ρ_{Fe} = ferromagnetic material density;

ρ_{Cu} = copper conductor density of coil;

k_{Fe} = core lamination factor.

Material cost

$$C = C'_{Cu} + C'_{Fe} = \sigma_{Fe}m_{Fe} + \sigma_{Cu}m_{Cu} \qquad (3\text{-}118)$$

where C_{Fe} = ferromagnetic material cost;

C_{Cu} = copper conductor cost of coil;

σ_{Fe} = ferromagnetic material cost per unit;

σ_{Cu} = copper conductor cost per unit of coil.

(3) Constraints

The constraints are the same to the static reliability optimization design of DC clapper electromagnetic system, i.e. the constraints of magnetic force [the same to the Formula (3-97)], temperature rise [the same to the Formula (3-88)], and magnetic flux density [the same to the Formula (3-89)] at design points and key points. In addition to performance constraints, the design variables should meet some constrains of geometrical dimensions, for example, every geometrical dimension should be more than zero, and some mutual relations.

2.6.2.3.2   DYNAMIC RELIABILITY OPTIMIZATION DESIGN OF AC ELECTROMAGNETIC SYSTEM

The method to establish the mathematic model is introduced as follows also in the case of double-E AC electromagnetic system.

(1)   Selection of Design Variables

The selection of variables is the same as the static reliability optimization design.

(2)   Objective Function

For dynamic reliability optimization design of electromagnetic system in electrical apparatus, the optimization objective should also be selected by taking into consideration energy and material saving on the premises of reliable electromagnetic system pull-in and release. Similar to static reliability optimization design, the material cost, power consumed by coil in rated thermal state, volume or weight are usually selected. Material cost is taken as the objective function in follows. The optimization objective should be selected by taking into consideration not only the static indexes, but also dynamic indexes. The contact colliding velocity v_1 and iron core colliding velocity v_2 are the key indexes that influence the working reliability and performance of AC electromagnetic system. In case of higher contact colliding velocity v_1 and iron core colliding velocity v_2, the colliding energy will increase, which will influence the mechanical life of electromagnetic system, aggravate contact bounce which increases contact wearing, and reduces contact electric life, consequently decreasing the performance indexes of electrical apparatus. Moreover, contact bounce may cause alternative connection and disconnection of the main circuit, and even contact splicing, reducing the contactor working reliability.

As to AC electromagnetic system, the dynamic process varies with the closing phase angle of coil voltage. So the contact colliding velocity v_1 and core colliding velocity v_2 will be different too. Therefore, to compare the performance of two electromagnetic systems, the mean colliding velocity v_{1pj} and the mean core colliding velocity v_{2pj} at every closing phase angle should be taken into consideration. Because the closing phase angle is random, it can be considered that the closing opportunities at every phase angle are equal. Therefore, we can obtain the mean colliding velocity

$$v_{1pj} = \frac{1}{180} \int_0^{180} v_1 d\theta \tag{3-117}$$

$$v_{2pj} = \frac{1}{180} \int_0^{180} v_2 d\theta \tag{3-118}$$

During dynamic reliability optimization design of AC electromagnetic system, v_{1pj} and v_{2pj} can also be taken as the objective functions, so the problem belongs to the multi-objective optimization problem.

As to multi-objective optimization problem, the minimization of certain sub-objective will worsen the optimal value of other sub-objectives because of the mutual contradiction between sub-objectives. For multi-objective optimization design, there are the following decision-making approaches generally:

Weighted factor approach: appoint weight to every sub-objective, and unify them into single-objective function. Non-inferior set will be generated when these weights are changed by parameters.

Coordination curve approach: when the optimal values of sub-objectives are contradictory, in order to let the sub-objective of certain relative larger value (in case the minimum value is optimization) reach reasonable value, we should coordinate the optimal values of sub-objective functions at the cost of increasing the other sub-objectives, i.e. by mutual concession, to finally obtain a generally satisfactory scheme.

Primary objective approach: first order the sub-objectives according to their importance in design by positive sequence, and then obtain the constrained optimal value of every sub-objective function in turn. During solution, give appropriate estimated optimal value to the other sub-objective functions as the decision-maker considered in preliminary design, just treat them as assistant constraints. Finally, obtain the relative optimal or satisfactory solution that can be accepted for the overall design.

Game-theoretic approach: consider each sub-object as a player, i.e. correspond each player to a sub-objective function to be optimized. The players always expect to select mutually beneficial strategy (eclectic solution) during negotiation, i.e. all of them are farthest from the bad pay off (objective function value).

The formation of new objective function by weighted factor approach and game-theoretic approach are introduced in detail as follows:

1) Weighted Factor Approach

By choosing different weighted factors according to the their importance of sub-objectives, magnitude and dimension, we can obtain the following new objective function:

$$\min \ f(X) = \alpha_1 (w_1 + w_2 + \alpha_0 w_3) + \alpha_2 v_{1pj} + \alpha_3 v_{2pj} \qquad (3\text{-}119)$$

where, w_1 is the movable iron core weight, w_2 is the static iron core weight, w_3 is the coil copper weight, v_{1pj} is the mean moving velocity of movable contact at collision between movable and static contacts, v_{2pj} is the mean moving velocity of movable iron core at collision between movable and static iron cores, α_0 is copper/iron price ratio, α_1, α_2, α_3 are respectively weighted parameters to allow optimization of all sub-objectives.

2) Game-theoretic Approach

First optimize every sub-objective as single objective to obtain every extreme point and corresponding sub-objective function value, as shown in Table 3-13.

Table 3- 13 Sub-objective function values at extreme points

Extreme point	f_c	$f_{v_{1pj}}$	$f_{v_{2pj}}$
X_c	$f_c(X_c)$	$f_{v_{1pj}}(X_c)$	$f_{v_{2pj}}(X_c)$
$X_{v_{1pj}}$	$f_c(X_{v_{1pj}})$	$f_{v_{1pj}}(X_{v_{1pj}})$	$f_{v_{2pj}}(X_{v_{1pj}})$
$X_{v_{2pj}}$	$f_c(X_{v_{2pj}})$	$f_{v_{1pj}}(X_{v_{2pj}})$	$f_{v_{2pj}}(X_{v_{2pj}})$

In Table 3-13, f_c, $f_{v_{1pj}}$, $f_{v_{2pj}}$ are respectively the objective function values of price, mean contact colliding velocity and mean iron core colliding velocity, and X_c, $X_{v_{1pj}}$, $X_{v_{2pj}}$ respectively are the extreme points obtained by optimization in case the price, mean contact colliding velocity and mean iron core colliding velocity are as single objective functions.

Suppose f_{cw} is the maximum among $f_c(X_c)$, $f_c(X_{v_{1pj}})$, $f_c(X_{v_{2pj}})$, and $f_{v_{1pj}w}$ is the maximum among $f_{v_{1pj}}(X_c)$, $f_{v_{1pj}}(X_{v_{1pj}})$, $f_{v_{1pj}}(X_{v_{2pj}})$, and $f_{v_{2pj}w}$ is the maximum among $f_{v_{2pj}}(X_c)$, $f_{v_{2pj}}(X_{v_{1pj}})$, $f_{v_{2pj}}(X_{v_{2pj}})$.

According to game theory, the new objective function is

$$\max f(X) = (f_{cw} - f_c)^{w_1} (f_{v_{1pj}w} - f_{v_{1pj}})^{w_2} (f_{v_{2pj}w} - f_{v_{2pj}})^{w_3} \qquad (3\text{-}120)$$

where w_1, w_2, w_3 are respectively the relative importance factor of each sub-objective function.

Formula（3-120）is the maximization problem of objective function, which can be transformed to minimization problem simply by multiplying the objective function by the value -1.

(3) Constraints

The magnetic force and motion characteristics of AC electromagnetic system are not only related to time but also the closing phase angle of coil voltage. It is obvious that the dynamic characteristics are very complex, especially the matching between characteristics of dynamic magnetic force and reaction force, but the requirement is explicit for matching. When the coil is suddenly energized with $85\% \sim 110\% U_e$, the contactor should be capable of reliable operating at every closing phase angle, i.e. its movable iron core should be attracted smoothly and thoroughly in constant closed state, but only released when coil is de-energized or the voltage drops to release value. It can be known from dynamic characteristic analysis that if the velocity is more than zero during the pull-in process, it means the movable core will not stagnate or be blocked midway. More relevantly, if there is no obvious phenomenon that the velocity is less than zero during pull-in process, it means the movable core will not stagnate or be blocked midway.

The so-called no obvious phenomenon that the velocity is less than zero refers to the phenomenon of AC electromagnetic system at some closing phase angle of coil voltage that the armature moving velocity is less than zero after the movable part is actuated during the moving stage of pull-in. This is caused by the slow increasing of the magnetic force less than the reaction force rapidly and insufficiency of the kinetic energy accumulation of the moving part. However, this phenomenon is temporary, and will not lead to operate failure or unsmooth motion of electromagnetic system. In other words, the armature of movable part is still capable of reliable, smooth and full pull-in within a period of time.

Therefore, velocity characteristic is a necessary constraint in the closing process. Specific constrains are as follows:

1)   The moving velocity of movable iron core should be more than zero after the normal close auxiliary contact is opened at every phase angle of coil voltage;

2)   The minimum magnetic force at closed position should be greater than the reaction force;

3)   The maximum coil temperature rise should not exceed the allowable temperature rise;

4)   The electromagnetic system action time should be less than the specified time;

5)   Constraints of optimization variable geometric dimensions.

(4)   Mathematic model of reliability optimization design of AC electromagnetic system for different new objective function

Because the formed new objective functions are different, the mathematic models of reliability optimization design of AC electromagnetic system are accordingly different.

1) Mathematic model of reliability optimization design of AC electromagnetic system with the objective of the Formula（3-119） is shown as the Formula （3-121）.

$$\min f(X) = \alpha_1(w_1 + w_2 + \alpha_0 w_3) + \alpha_2 v_{1pj} + \alpha_3 v_{2pj}$$

$$\text{s.t. } v\Big|_{\substack{x>x_1 \\ U=U_R}} \geq 0$$

$$F_{\min}\Big|_{\substack{\delta=\delta_{\min} \\ U=U_R}} > F_f \tag{3-121}$$

$$t_{cd} \leq t_{cdm}$$

$$\tau_{\max}\Big|_{\substack{\delta=\delta_{\min} \\ U=K_u U_e}} \leq \tau_m$$

$$X_{i\min} \leq X_i \leq X_{i\max}$$

Constraints of other geometric dimensions

where x is the displacement; x_1 is the overtravel of the normal close auxiliary contact; τ_m is the maximum allowable temperature rise; t_{cdm} is the maximum allowable pull-in time; F_f is the reaction force at closed position; $F_{\min}$ is the minimum magnetic force at closed position; and X_i, $X_{i\min}$, $X_{i\max}$ are the ith design variable and its acceptable minimum and maximum.

The constraints should be treated specifically according to the actual situation. For example, the first constraint and the constraints of variable geometric dimensions must be met. Otherwise, it may lead to computing program error. Therefore, the design sets that do not meet these constraints should be automatically rejected. However, this practice will limit the capacity of exploring the whole design space, and consequently trap the optimization in local extreme points. So the design sets should not be automatically rejected when they do not meet other performance constraints. These constraints should be alternatively transformed as a part of the objective function. Thus the formula can be transformed to

$$\min \quad f(X)(1 + R_1\varphi_1(F(X)) + R_2\varphi_2(\tau(X)) + R_3\varphi_3(t(X)))$$

$$\text{s.t. } v\Big|_{\substack{x>x_1 \\ U=U_R}} \geq 0 \tag{3-122}$$

$$X_{i\min} \leq X_i \leq X_{i\max}$$

Constraints of other geometric dimensions

where R_1, R_2, and R_3 are respectively penalty factors in every penalty functions that are as follows:

$$\varphi_1(F(X)) = \begin{cases} 0 & \left(F_{\min}\Big|_{\substack{\delta=\delta_{\min} \\ U=U_R}} > F_f\right) \\ \\ F_f - F_{\min}\Big|_{\substack{\delta=\delta_{\min} \\ U=U_R}} & \left(F_{\min}\Big|_{\substack{\delta=\delta_{\min} \\ U=U_R}} \le F_f\right) \end{cases}$$

(3-123)

$$\varphi_2(\tau(X)) = \begin{cases} 0 & \tau_{\max}\Big|_{\substack{\delta=\delta_{\min} \\ U=K_U U_e}} \le \tau_m \\ \\ \tau_{\max}\Big|_{\substack{\delta=\delta_{\min} \\ U=K_U U_e}} - \tau_m & \tau_{\max}\Big|_{\substack{\delta=\delta_{\min} \\ U=K_U U_e}} > \tau_m \end{cases}$$

(3-124)

$$\varphi_3(t(X)) = \begin{cases} 0 & t_{cd} \le t_{cdm} \\ t_{cd} - t_{cdm} & t_{cd} > t_{cdm} \end{cases}$$

(3-125)

2) Mathematic model of reliability optimization design of AC electromagnetic system with object of the Formula（3-120）is shown as the Formula（3-126）.

$$\begin{cases} \min & -f(X)(1 + R_1\varphi_1(F(X)) + R_2\varphi_2(\tau(X)) + R_3\varphi_3(t(X))) \\ \\ \text{s.t.} & v\Big|_{\substack{x>x_1 \\ U=U_R}} \ge 0 \\ \\ & X_{i\min} \le X_i \le X_{i\max} \\ \\ & \text{Every multiplied factor in } f(X) \text{ should be more than zero} \\ & \text{Constraints of other geometric dimensions} \end{cases}$$

(3-126)

The meanings of the variables in the formula are the same as the above.

2.6.2.4   FUZZY RELIABILITY OPTIMIZATION DESIGN OF ELECTROMAGNETIC SYSTEM IN ELECTRICAL APPARATUS

In the design of electromagnetic system for electrical apparatus, the reliability design that is mathematically based on probability theory and mathematic statistics is called general reliability design where the random phenomena that exist is considered. However, besides random phenomena, there is also a great deal of fuzzy phenomena that can exist in the design of electromagnetic system in electrical apparatus. The general reliability design theory is incapable of handling the fuzzy phenomena, which claims method to apply fuzzy mathematics in the general reliability design theory. This method combining the fuzzy mathematics and general reliability design theory is called fuzzy reliability design method, which allows the handling of both random and fuzzy phenomena.

Fuzzy mathematics expands the application scope of mathematics from the category of precision phenomena (either-or) to the category of fuzzy phenomena (both A and B). Fuzzy random reliability design (abbreviated as fuzzy reliability design) is a design theory and method based on fuzzy mathematics and general reliability design theory (probability theory and mathematic statistics), taking into consideration both fuzzy and random phenomena.

2.6.2.4.1 FUZZY RELIABILITY DESIGN OF ELECTROMAGNETIC SYSTEM IN ELECTRICAL APPARATUS

(1) Rule of Fuzzy Reliability Design
In general reliability design, the qualification of electromagnetic system is denoted by $A=\{U_x \leq U_{xy}\}$, while the state of electromagnetic system is denoted by Ω, so its rule can be expressed by characteristic function $C_A(\Omega)$ as

$$C_A(\Omega) = \begin{cases} 1 & \Omega \in A \\ 0 & \Omega \notin A \end{cases} \qquad (3\text{-}127)$$

Considering there is a transitional process when the electromagnetic system changes from quality to un-quality, i.e. fuzziness. So the fuzzy event of electromagnetic system qualified is denoted by $\widetilde{A}$, and its qualified degree generally can be described by fuzzy event's membership function $\mu_{\widetilde{A}}(u)$. This fuzzy event can be denoted by $\widetilde{A} = \{U_x \leq U_{xy}\}$, and its membership function denoted as $\mu_{\widetilde{A}}(U_x \leq U_{xy})$. In case of U_x=187V and U_{xy}=187V, the electromagnetic system is qualified, i.e. $\mu_{\widetilde{A}}(187,187) = 1$. In case of U_x=187.1V and U_{xy}=187 V, the electromagnetic system is in the fuzzy state of both qualified and unqualified, and its membership function in this fuzzy state of qualified $\mu_{\widetilde{A}}(187,187)$ should be valued between 0 and 1. The fuzzy event's membership function $\widetilde{A} = \{U_x \leq U_{xy}\}$ can be established according to the actual situation. We call the design rule $U_x \stackrel{\sim}{\leq} U_{xy}$ fuzzy design rule of electromagnetic system (where "$\stackrel{\sim}{\leq}$" means approximate equal or less than).

In fuzzy design rule, qualified, unqualified and the transitional state are considered, which are more complying with the actual project.

(2) Basic Principle of Fuzzy Reliability Design

In fuzzy reliability design, its reliability refers to the probability of fuzzy event $\widetilde{A}$. Suppose load and strength are respectively y and x, then this fuzzy event of product qualified is denoted as $\dot{A} = \{y \, \widetilde{\le} \, x\}$, i.e.

$$R = P(y \, \widetilde{\le} \, x) \tag{3-128}$$

Its reliability can be calculated by the joint probability density function integral. Suppose the probability density function of strength is $f(x)$, and the probability density function of load is $g(y)$. Its load-strength interference model is shown as Figure 3-34a), in which x_0 is a certain strength value. The figure 3-34b) is the membership function of $y \, \widetilde{\le} \, x$, describing the transitional state during the occurrence of failure.

In Figure 3-34, we can obtain by general reliability theory that

$$P(y \le x_0) = \int_{-\infty}^{x_0} g(y)dy = \int_{-\infty}^{+\infty} g(y)C_A(y)dy$$

where $C_A(y)$ is the characteristic function of ordinary event A= $\{y \le x_0\}$;

By generalizing the above formula, we can obtain the probability of $y \, \widetilde{\le} \, x_0$:

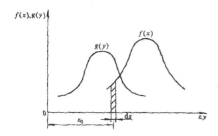 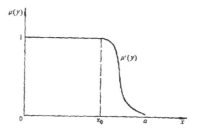

a.   Basic interference diagram   b.   Membership function

Figure 3-34   Fuzzy load-strength interference model

$$P(y \, \widetilde{\le} \, x_0) = \int_{-\infty}^{+\infty} g(y)\mu(y)dy \tag{3-129}$$

where $\mu(y)$ is the membership function of fuzzy event $\widetilde{A} = \{y \, \widetilde{\le} \, x_0\}$.

$$\mu(y) = \begin{cases} 1 & y \leq x_0 \\ \mu'(y) & x_0 \leq y < a \\ 0 & y \geq a \end{cases} \qquad (3\text{-}130)$$

where $\mu'(y)$ is used to describe fuzziness. In case $a=x_0$, it's the reliability design without taking fuzziness into consideration.

The probability of strength x in interval $(x_0\text{-}dx/2,\ x_0\text{+}dx/2)$

$$P(x_0 - \frac{dx}{2} \leq x \leq x_0 + \frac{dx}{2}) = f(x_0)dx$$

Suppose $(y \widetilde{\leq} x)$ and $(x_0 - \frac{dx}{2} \leq x \leq x_0 + \frac{dx}{2})$ are two independent events,

it can be known by multiplication theorem of probability that when strength x falls in interval $(x_0\text{-}dx/2,\ x_0\text{+}dx/2)$, the reliability is

$$dR = f(x_0)dx \int_{-\infty}^{+\infty} g(y)\mu(y)dy$$

Taking into consideration the range of all possible strength, the product reliability is

$$R = \int_{-\infty}^{+\infty} dR = \int_{-\infty}^{+\infty} f(x)[\int_{-\infty}^{+\infty} g(y)\mu(y)dy]dx$$

$$= \int_{-\infty}^{+\infty} f(x)\int_{-\infty}^{x_0} g(y)dy + \int_{-\infty}^{+\infty} f(x)[\int_{x_0}^{a} g(y)\mu'(y)dy]dx \qquad (3\text{-}131)$$

In Formula (3-131), the antecedent is the reliability formula calculated in general reliability design, and the consequent is the part added when fuzziness is taken into consideration.

2.6.2.4.2 CALCULATION FORMULA OF FUZZY RELIABILITY OF ELECTROMAGNETIC SYSTEM IN ELECTRICAL APPARATUS

As to electromagnetic system in certain electrical apparatus, its pull-in voltage U_x is a random variable, and its probability density function is expressed by $g(U_x)$; the allowable pull-in voltage is a given value U_{xy}, then it is a fuzzy event that the pull-in voltage is approximate equal to or less than the allowable pull-in voltage, i.e. $\widetilde{A} = \{U_x \widetilde{\leq} U_{xy}\}$. This fuzzy event's membership function is expressed by $\mu(U_x)$. The electromagnetic system pull-in reliability is:

$$R = P(U_x \tilde{\leq} U_{xy}) = \int_{-\infty}^{\infty} g(U_x)\mu(U_x)dU_x$$

$$= \int_{-\infty}^{a_1} g(U_x)dU_x + \int_{a_1}^{a_2} g(U_x)\mu'(U_x)dU_x \qquad (3\text{-}132)$$

where $\mu(U_x) = \begin{cases} 1 & U_x \leq a_1 \\ \mu'(U_x) & a_1 < U_x < a_2 \\ 0 & U_x \geq a_2 \end{cases}$

Its membership function can be determined by the accumulated data, previous experience and subjective judgment, with certain flexibility. Parameters a_1 and a_2 are the acceptable upper and lower limits of the strengths in the transitional process that are determined to describe the fuzzy event. a_1 is generally the given allowable pull-in voltage U_{xy}, and a_2 can be determined by augmented coefficient method according to experience, which is generally taken as ka_1. For mechanical fuzzy reliability design one can use $k=1.05$ to 1.3, and for electromagnetic system, $k=1.0$ to 1.06.

In case $k=1$, it is reliability design without taking fuzziness into consideration; and in case $k=1.06$, as to $U_e=220V$, $kU_{xy}=0.85kU_e=198.22V$.

While the minimum grid voltage is about 200V in general, so the range is reasonably selected.

The membership function can adopt both linear and normal form in fuzzy reliability design of electromagnetic system. Its calculation formula of reliability is discussed as follows:

In the case of normal form, membership function is shown in Figure 3-35

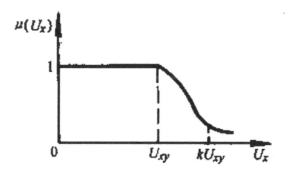

Figure 3-35 Normal membership function

i.e. $\mu(U_x) = \begin{cases} 1 & U_x \le U_{xy} \\ \exp[-\dfrac{(U_x - U_{xy})^2}{b^2}] & U_x > U_{xy} \end{cases}$ (3-133)

In case $U_x \sim N(\overline{U}_x, \sigma^2)$, the reliability is

$$R = P(U_x \overset{\sim}{\le} U_{xy}) = \int_{-\infty}^{\infty} g(U_x)\mu(U_x)dU_x$$

$$= \int_{-\infty}^{U_{xy}} g(U_x)dU_x + \int_{U_{xy}}^{\infty} \frac{1}{\sqrt{2\pi}\sigma} \exp(-\frac{(U_x - \overline{U}_x)^2}{2\sigma^2}) \exp(-\frac{(U_x - U_{xy})^2}{b^2})dU_x$$

Let $\overline{U}'_x = \dfrac{b^2\overline{U}_x + 2\sigma^2 U_{xy}}{b^2 + 2\sigma^2}$, $\sigma' = \dfrac{\sigma b}{\sqrt{2\sigma^2 + b^2}}$, then

$$R = \Phi(\frac{U_{xy} - \overline{U}_x}{\sigma}) + \frac{b}{\sqrt{2\sigma^2 + b^2}} \exp[-\frac{(U_{xy} - \overline{U}_x)^2}{(2\sigma^2 + b^2)}]$$

$$\times \int_{U_{xy}}^{\infty} \frac{1}{\sqrt{2\pi}\sigma'} \exp\left[-\frac{(U_x - \overline{U}'_x)^2}{2\sigma'^2}\right]dU_x$$

Let $s = \dfrac{U_x - \overline{U}'_x}{\sigma'}$, then $dU_x = ds$, so

$$R = \Phi(\frac{U_{xy} - \overline{U}_x}{\sigma}) + \frac{b}{\sqrt{2\sigma^2 + b^2}} \exp[-\frac{(U_{xy} - \overline{U}_x)^2}{(2\sigma^2 + b^2)}]$$

$$\times \int_{\frac{U_{xy} - \overline{U}'_x}{\sigma'}}^{\infty} \frac{1}{\sqrt{2\pi}} \exp\left[-\frac{s^2}{2}\right]ds$$

$$= \Phi(\frac{U_{xy} - \overline{U}_x}{\sigma}) + \frac{b}{\sqrt{2\sigma^2 + b^2}} \exp[-\frac{(U_{xy} - \overline{U}_x)^2}{(2\sigma^2 + b^2)}] \times \Phi(-\frac{U_{xy} - \overline{U}'_x}{\sigma'})$$

$$= \Phi(\frac{U_{xy} - \overline{U}_x}{\sigma}) + \frac{b}{\sqrt{2\sigma^2 + b^2}} \exp[-\frac{(U_{xy} - \overline{U}_x)^2}{(2\sigma^2 + b^2)}]$$

$$\times \Phi(-\frac{b}{\sqrt{2\sigma^2 + b^2}} \frac{U_{xy} - \overline{U}_x}{\sigma})$$ (3-134)

The Formula（3-134） involves the selection of the value of b. By taking into consideration that for linear membership function, a_1 is the allowable pull-in voltage U_{xy}, and a_2 is ka_1, b can be determined on "3σ" principle for normal membership function, i.e. $b = \dfrac{a_2 - a_1}{3} = \dfrac{(k - 1)a_1}{3}$.

In that case, k=1.05, and b=0.017a_1=0.017U_{xy}.

2.6.2.4.3 FUZZY RELIABILITY OPTIMIZATION DESIGN OF ELECTROMAGNETIC SYSTEM IN ELECTRICAL APPARATUS

The fuzzy reliability optimization design method of electromagnetic system in electrical apparatus is similar to the general reliability optimization design method. First we should establish the mathematical model of fuzzy reliability optimization design, select optimization algorithm and conduct optimization calculation to obtain the optimization scheme. What is different from the general reliability optimization design is the mathematical model, and its establishment method is explained as follows.

(1)   Take fuzzy pull-in reliability R as objective function

When taking fuzzy pull-in reliability R as the optimization objective, the following problems should be solved:

1) Calculation of Average Pull-in Voltage and Standard Deviation

In case the objective function is fuzzy pull-in reliability R, the fuzzy pull-in reliability R of the design scheme that corresponds to each set of design variables should be calculated during the process of optimization. The average pull-in voltage and standard deviation of this design scheme should be first calculated for the calculation of fuzzy pull-in reliability R.

2) Selection of Fuzzy Membership Function

The fuzzy membership function can be selected according to the actual situation and previous experience. It can be the linear membership function or normal membership function. No matter what kind of membership function, the value k should be determined in terms of normal membership function. After value k is determined, value b should be calculated, and then the expression of membership function obtained by the Formula （3-133）.

3) Calculation of Its Fuzzy Pull-in Reliability R

In terms of normal membership function, the fuzzy pull-in reliability R can be calculated by the Formula （3-134）.

4) Constraints

The constraints are the same as that in general reliability optimization design of electromagnetic system in electrical apparatus.

5) Selection of Optimization Algorithm

The selection of optimization algorithm is the same as that in the general reliability optimization design. The global optimization algorithm can be selected, such as Simulated Annealing method, Genetic Algorithm, and Tabu Algorithm, etc..

(2)   With Constraint of Fuzzy Pull-in Reliability R

During the optimization with constraint of fuzzy pull-in reliability R, the fuzzy reliable pull-in voltage U_R should be calculated first. In terms of normal membership function, this pull-in voltage is the average pull-in voltage in Formula (3-134). The deviation in Formula (3-134) can be calculated by the product of its average pull-in voltage value and variation coefficient. The variation coefficient of pull-in voltage can be selected according to the previous experience, or refer to the data on that of the products with the similar structure.

It can be seen from the Formula (3-134) that the relation between reliability R and average pull-in voltage $\overline{U}_x$ is very complex, and reliability R cannot be obtained by analytic method, but the method of single variable optimization can be adopted. Because R is monotonic decreasing with $\overline{U}_x$, i.e. the larger the $\overline{U}_x$, the smaller the R. The method of bisection can be adopted to calculate the fuzzy reliable pull-in voltage U_R.

4

RELIABILITY OF CONTROL
RELAY

Control relay is a basic electrical apparatus element with large amount and extensive range, widely applied in the sectors of machine, electronics, aerospace and aviation, railways, post and telecommunication, and electric power, etc. A large equipment or system usually adopts many relays, in order to ensure higher reliability of equipment and system, the control relay, as one of the main basic elements in the equipment or system, shall have high reliability, so the reliability of control relay has received widespread attention of international researchers, and the relay standards with reliability indices have been prepared in many countries (such as the U.S. and Japan, etc.). In 2011, IEC issued the IEC61810-2 "Electromechanical elementary relays-Part2: Reliability".

1 RELIABILITY INDEX

In order to unify the assessment method of relay reliability, and to further promote the relay reliability development in China according to the national standards preparation plan issued by the State Bureau of Technical Supervision of China, the national standard GB/T 15510-1995 *General Rules for Reliability Test of Electromagnetic Relay used in Control Circuits* was prepared by the former Beijing Electrical Polytechnic Technology and Economic Institute under the Ministry of Mechanical Industry and the Hebei University of Technology. In this national standard, it's specified that the reliability grades of control relay should be classified according to its failure rate level, with the name, symbol of failure rate grade, and the maximum failure rate of each grade shown in Table 4-1.

Table 4-1 Failure rate grade symbol and maximum failure rate of control relay

Name of failure rate grade	Symbol of failure rate grade	Maximum failure rate λ_{max}(1/10 times)
Subgrade V	YW	3×10^{-5}
Grade V	W	1×10^{-5}
Grade VI	L	1×10^{-6}
Grade VII	Q	1×10^{-7}

In IEC 61810-2, the loads of relay contact are classified into cc1 (refers to the low load without arc) and cc2 (refers to the high load which can generate arc). For the relay with contact load cc1, its failure mode can be considered as random failure, and its life distribution can be considered as subject to exponential distribution, so the failure rate λ can be used as the characteristic quantity of its reliability, and the relay reliability grade can be classified according to the maximum failure rate value λ_{max}, its method is similar to Table 4-1; For the relay with contact load cc2, its failure mode is wear-out failure, and its life distribution can be considered as subject to Weibull distribution, so the reliability can be used as the characteristic quantity of its reliability, and the relay reliability grade can be classified according to the reliability value at the rated life. The name of the reliability grade and reliability of rated life are shown in Table 4-2.

Table 4-2 Reliability grade of relay with contact load cc2

Name of Reliability Grade	Reliability $R(T_e)$ at Rated Life T_e
Grade I	0.95
Grade II	0.9
Grade III	0.85

2   TEST REQUIREMENTS

2.1 ENVIRONMENTAL CONDITIONS

(1) In general, the test should be conducted under the standard atmospheric conditions which were specified in GB 2423 *Basic Environmental Testing Procedures for Electric and Electronic Products*, i.e.

Temperature   15 to 35°C
Relative Humidity   45% to 75%
Atmospheric pressure   86 to 106kPa

The test sample should be placed in the standard atmospheric conditions for enough time (not less than (8h), for making the test sample reach thermal equilibrium.

(2) The test environment should be prevented from dust and other pollutants.

2.2 INSTALLATION CONDITIONS

(1) The test sample should be installed in the place of normal application.

(2) The test sample should be installed in the place without remarkable impact and vibration.

(3) The inclination between the installing surface and vertical surface of the test sample should be consistent with the product standard.

2.3 CONDITIONS OF POWER SUPPLY

(1) The AC power supply should be sine-wave power with 50Hz frequency, and the allowable deviations:

    1）Waveform distortion factor should not exceed 5%;

    2）Frequency deviation of ±5%.

(2) The DC power supply may adopt generator, accumulator or stabilized voltage power supply, or 3-phase full-wave rectifying power supply in the case the product performances are not affected in the test, but its ripple component should meet the following regulation: the ratio between the peak/valley difference and the DC component should not exceed 6%.

(3) When the contact is connected to the load during the test, the voltage fluctuation of test power supply relative to the no-load voltage should not exceed 5%.

2.4 LOAD CONDITIONS

2.4.1 RELAY WITH CONTACT LOAD CC1

(1) Load power supply can adopt DC or AC power supply, but the DC power supply is recommended in general.

(2) The load can be resistive load, inductive load, capacitive load or nonlinear load, but the resistive load is recommended in general (AC $\cos\varphi=0.9\sim1.0$; DC $L/R\leq1ms$).

(3) In general, the power voltage U_N in contact circuit should be 10V, or the minimum rated DC voltage specified in the product standard.

(4) In general, the load current I_C in contact circuit should be 10mA or 100mA.

2.4.2 RELAY WITH CONTACT LOAD CC2

Contact circuit power voltage should be the rated product voltage, and contact circuit load current should be the rated current.

2.5 EXCITATION CONDITIONS

(1) In the test, the test sample should be excited by the rated input energizing quantity.

(2) Number of cycles per hour: the number of cycles per hour in the test should not be less than the rated value specified in the product standard; in order to shorten test time, in the case that the normal test sample action and release are not affected, the number of cycles per hour of test sample can be more than the rated value in the product standard, selected from 6, 30, 600, 1200, 1800, 3600, 7200, 12000, 18000 and 36000.

(3) The load ratio (load factor) should be selected from the following recommended values: 15%, 25%, 33%, 40%, 50% and 60%.

3 TEST METHOD

3.1 GENERAL RULES FOR RELIABILITY TEST

3.1.1 TYPES OF RELIABILITY TEST

Reliability tests can be divided into reliability determination test and reliability verification test.

3.1.1.1 RELIABILITY DETERMINATION TEST

Reliability determination test refers to the test to determine the characteristic value of product reliability, and is generally used to provide the reliability data, applicable to the product without quantitative specification of reliability requirements. By reliability determination test, the product reliability level can be evaluated.

3.1.1.2 RELIABILITY VERIFICATION TEST

Reliability verification test refers to the test to verify whether the characteristic value of product reliability complies with the specified reliability requirement, it is generally one of the conditions for product acceptance by the ordering party.

3.1.2 DETERMINATION OF TEST LOCATION

The advantages of lab test are that the test conditions can be limited and controlled, and the test results are comparable. In addition, in lab test, the tested product performance can be better monitored and its failure can be better displayed. In many cases, the lab test conditions can be designed accurately according to the maximum application conditions.

The advantages of field test are that the provided test results are more realistic, the required test devices are fewer, the test cost is lower than the cost for corresponding lab test, and the tested product can work in normal conditions. The drawback of field test is that the test cannot be conducted in restrictively controlled conditions.

3.1.3 DETERMINATION OF TEST PLANS

In the case of constant failure rate, the following two types of test plans can be adopted.

(1) Truncation sequential test plan: during the test, monitor the product in succession or short intervals, and compare the accumulated test time and failure number with the specified criteria to determine whether to accept, reject, or further test the product.

(2) Time or failure curtailed test plan: during the test, monitor the product in succession or short intervals, and in the case the accumulated test time reaches the predetermined test time, and the failure number is less than the predetermined failure number, it's judged acceptable; in the case the accumulated test time fails to reach the predetermined test time, while the

failure number reaches the predetermined failure number, it's judged rejectable.

The advantages and disadvantages of the above two test plans in terms of economy and management are as follows:

Advantages of truncation sequential test plan:
(1) Minimum average failure number required by judgment;
(2) Minimum average accumulated test time required by judgment.

Disadvantages of truncation sequential test plan:
(1) The variation amplitude of failure number and the relevant test sample cost are larger than those of similar time or failure curtailed test plan, leading to problems in arranging the test samples, test device, manpower and other management problems.
(2) The maximum accumulated test time and failure number may exceed those of corresponding time or failure curtailed test plan.

Advantages of time or failure curtailed test plan:
(1) The maximum accumulated test time is fixed, so the maximum required quantity of test devices and manpower can be determined before the test.
(2) The maximum quantity of test samples can be determined before the test.

Disadvantages of time or failure curtailed test plan:
(1) Both average failure number and average accumulated test time will exceed those of corresponding truncation sequential test plan.
(2) No matter of poor or good product quality, the judgment can be made only in the case the predetermined accumulated test time or failure number is reached, while this judgment is made faster in corresponding truncation sequential test plan.

3.2 PREPARATION OF TEST SAMPLE

For the relay with contact load cc1, in order to meet assumption that the product life is subject to exponential distribution, the products failed in early-stage should be eliminated by screening method, so the test samples should be randomly sampled from the qualified products produced by batch production under the stable process conditions. To prevent excessive test complexity, it's recommended the screening should be conducted at normal temperature (15~35°C), for 5000 times; the excitation condition, contact circuit power voltage U_N and contact circuit load current are the same with the test requirements mentioned above.

3.3 INSPECTION OF TEST SAMPLE

3.3.1 INSPECT BEFORE THE TEST

Conduct open package inspection for the test samples before the test to check whether the components of the test samples are damaged or broken due to transportation, eliminate the test sample with damaged part or component, replace them by new ones according to specified requirements. The eliminated test samples should not be counted into the relevant failure number r.

3.3.2 INSPECTION DURING THE TEST

In general, during 40% of "on" time and during 40% "off" time at each cycle, all contacts, the contact voltage drop at closed contact and voltage between open contacts of the test sample should be detected. The test sample should not be sorted out or adjusted during the test.

3.3.3 INSPECTION AFTER THE TEST

In general, all test samples without failure after the test should be inspected according to the following items:
1) Appearance inspection;
2) Operating voltage;
3) Release voltage;
4) Contact resistance;
5) Insulation resistance;
6) Dielectric voltage;
7) Pull-in time;
8) Release time;
9) Bounce time;
10) Coil Resistance

3.4 FAILURE CRITERIA

When any one of the following cases occurs, the test sample is considered to be failure.

(1) Contact voltage drop U_j exceeds the following limit value U_{jm}.

    1)   In the case of rated load current, the limit value of contact voltage drop U_{jm} is 5% or 10% of power voltage of contact circuit U_N.

    2)   n the case of 10mA or 100mA load current, the limit value of contact voltage drop U_{jm} is as shown in Table 4-3.

(2) The voltage U_C between open contacts is lower than the limit value U_{cx}, in general, U_{cx} should be 90% of power voltage of contact circuit.

(3)   Contact welding or sticking.

(4) The arcing time of contact exceeds 0.1s.

(5) Relay does not take action when coil is energized.

(6) Relay does not return after coil is de-energized.

(7) The sample part is seriously damaged, and the connecting wire and parts loosen.

(8) The test samples should be detection after test, the test result of an items should comply with the specifications of product standard.

Table 4-3 Limit value of contact voltage drop U_{jm} at contact

Contact Circuit Load Current I_C(mA)	Limit Value of Contact Voltage Drop U_{jm} (V) at Contact
10	0.1
100	0.5

4 RELIABILITY TEST OF RELAY WITH CONTACT LOAD CC1

Reliability verification test is generally adopted for the relay with load type of cc1.

4.1 SAMPLING PLAN OF RELIABILITY VERIFICATION TEST

The reliability verification test of relay should be conducted in lab, which is generally recommended to adopt the time or failure curtailed test plan. This can be divided into 3 parts: classification test, maintenance test, and upgrade test.

1. Classification test refers to the primary test conducted to determine the product's failure rate grade, or the test conducted to re-determine its failure rate grade after the maintenance test or upgrade test of certain failure rate grade is failed.
2. Maintenance test refers to the test conducted to prove the product failure rate grade is still not lower than the failure rate grade determined after the classification test or upgrade test.
3. Upgrade test refers to the test conducted to prove the product failure rate grade is higher than the originally determined failure rate grade.

In case of life subject to single coefficient distribution, the truncation time T_c in reliability verification test plan is as follows.

$$T_c = \frac{\chi^2_{1-\beta}(2A_c + 2)}{2\lambda_1} \tag{4-1}$$

In which, the relationship between $2\lambda_1 T_c$ and β is as shown in Figure 4-1.

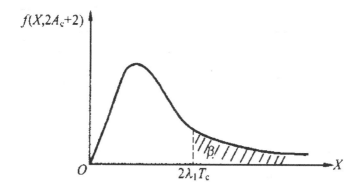

Figure 4-1 Relation between $2\lambda_1 T$ and β

$2\lambda_1 T_c$ equals to the quantile $\chi^2_{1-\beta}(2A_c + 2)$ under $1-\beta$ distributed by χ^2 with freedom of $2A_c + 2$.

For different value A_c, the corresponding value T_c can be obtained to determine the sampling plan. For classification test and upgrade test, the confidence is 0.9, with sampling plan as shown in Table 4-4. For maintenance test, the confidence is 0.6, with sampling plan as shown in Table 4-5.

Table 4-4 Sampling plan for classification test and upgrade test

Allowable Number of Failure A_C / Truncation time T_C (10^6 times) / Failure Rate Grade	0	1	2	3	4	5	6	7	8	9
YW	0.768	1.30	1.77	2.23	2.66	3.09	3.51	3.92	4.33	4.74
W	2.30	3.89	5.32	6.68	7.99	9.27	10.53	11.77	13.0	14.21
L	23.0	38.9	53.2	66.8	79.9	92.7	105.3	117.7	130.0	142.1
Q	230	389	532	668	799	927	1053	1177	1300	1421

Table 4-5 Sampling plan for maintenance test

Failure Rate Grade	Max. Maintenance Cycle (Month)	Truncation Time T_C (10^6 times)									
		$A_C=0$	$A_C=1$	$A_C=2$	$A_C=3$	$A_C=4$	$A_C=5$	$A_C=6$	$A_C=7$	$A_C=8$	$A_C=9$
YW	6	0.308	0.673	1.03	1.39	1.75	2.10	2.45	2.80	3.15	3.50
W	6	0.916	2.02	3.10	4.18	5.25	6.30	7.35	8.40	9.44	10.5
L	12	9.16	20.2	31.0	41.8	52.5	63.0	73.5	84.0	94.4	105
Q	24	91.6	202	310	418	525	630	735	840	944	1050

4.2 PROCEDURE OF RELIABILITY VERIFICATION TEST

4.2.1 CLASSIFICATION TEST

Procedure of classification test:

(1) To select the failure rate grade, YW or W should be generally selected for the first classification test.

(2) To select the allowable number failure of A_C and truncation failure number r_c $(r_c = A_c + 1)$, and A_C is recommended within 2~5, but not A_C=0.

(3) According to the selected failure rate grade and A_C, find the truncation time T_c from Table 4-4.

(4) Select the test deadline t_z of test sample which should not exceed the electrical endurance specified in the product standard, but shall not be less than 10^5 times.

(5) According to T_C, A_C and t_z, determine the test sample number n by the following equation, i.e.

$$n = \frac{T_C}{t_z} + A_C \qquad (4\text{-}2)$$

It should be noticed that the test sample number n should not be less than 10 times in general.

(6) Sample n test samples from the conforming products after screening that are produced by mass production, and the product number for sampling should not be less than 10 times that of the samples n.

(7) Conduct test and detection complying with the test method given in this chapter.

(8) To make statistics the relevant failure number r, and the relevant test time (failure occurrence time) of each failed test sample; for the relevant failure test samples which were detected after test, the relevant test time should be calculated according to t_z.

(9) To make statistics the accumulated relevant test time T.

(10) Judgment of test results: in the case the relevant failure number r does not reach the truncation failure number r_C (i.e. $r \le A_C$), while the accumulated relevant test time T reaches or exceed truncation time T_C, it will be judged as test qualified(acceptance); in the case the accumulated relevant test time T does not reach the truncation time T_C, while the relevant failure number r reaches the truncation failure number r_C ($r > A_C$), it will be judged as test unqualified(rejection).

4.2.2 MAINTENANCE TEST

In general, for the products that have passed classification test, the maintenance test of the same grade should be conducted according to the maintenance cycle given in Table 4-5, and with the following procedures:

(1) Select the allowable failure number A_C.

(2) Inquire the truncation time T_C in Table 4-5 according to the tested conforming failure rate grade, and the selected allowable failure number of the product.

(3) Select the test deadline t_z of the test sample (the method is same with that of the classification test).

(4) Determine the sample number n (the method is same with that of the classification test).

(5) Sample the test samples (the method is same with that of the classification test).

(6) Conduct tests and inspections complying with the test method given in this chapter.

(7) Perform statistics of the relevant failure number r, and the relevant test time of each failed test sample (the method is same with that of the classification test).

(8) Perform statistics of the accumulated relevant test time T.

(9) Judge the test result (the method is same with that of the classification test).

(10) If the maintenance test is conforming, conduct the next maintenance test according to the specified maintenance cycle. If the maintenance test is nonconforming, re-conduct the classification test to determine its failure rate grade.

(11) When re-determining the failure rate grade, accumulate all product test data from the primary classification test (including the nonconforming maintenance test data), and determine the product failure rate grade by Table 4-4 according to the accumulated relevant failure number and accumulated relevant test time.

4.2.3 UPGRADE TEST

Upgrade test can be further conducted on the conforming products after classification test. The upgrade test data can be obtained by conducting the prolonged test on the test sample of classification test and maintenance test, and the test on the test sample input for the upgrade test. The upgrade test should be conducted by the following procedure.

(1) To select the failure rate grade to be upgraded (which is generally one grade higher than the originally determined one).

(2) To select the allowable failure number A_C.

(3) According to the selected failure rate grade and allowable failure number, inquire the truncation time T_C in Table 4-4.

(4) According to T_C, determine the prolonged test time, and the test sample number and test time of upgrade test.

(5) Sample the test samples (the method is same with that of the classification test).

(6) Conduct test and detection according to regulations.

(7) To make statistics the relevant failure number r and accumulated relevant test time T。

(8) Judge the test result (the method is same with that of the classification test).

(9) If the upgrade test is conforming, conduct the maintenance test of the same grade according to the specified maintenance cycle; if the upgrade test is nonconforming, re-conduct the classification test to determine its failure rate grade.

(10) When re-determining the failure rate grade, accumulate all test data of the product, and determine the product failure rate grade by Table 4-4 according to the accumulated relevant failure number and accumulated relevant test time.

5   RELIABILITY TEST OF RELAY WITH CONTACT LOAD CC2

Reliability test of relay with contact load cc2 should adopt reliabiity determination test in the lab.   It's recommended that the fixed-number truncation test be adopted.

5.1 PROCEDURE OF RELIABILITY DETERMINATION TEST

a) Detemine test sample number n generally within the range of $10\sim20$;

b) Randomly sample n test samples from the conforming products that are produced by mass production; the product number for sampling should not be less than 10 times of the test sample number n;

c) Conduct test and detection according to the method given in this chapter, and judge whether the test sample failure is in accordance with the provisions of 3.4 in this chapter;

d) When more than 7 to 10 test samples fail, stop the test;

e) Assess the reliability grade according to the result of reliability determinatin test.

5.2 RELIABILITY GRADE DETERMINATION METHOD OF RELAY WITH CONTACT LOAD CC2

(1) Statistically analyze each failed test sample's relevant test time (failure occurrence time) t_i; For the relevant failed test samples detected after test, the relevant test times should be calculated according to the times at the end of the test.

(2) In case $n>20$, $F(t_i)=i/n$; in case $n\leq20$, $F(t_i)=(i-0.3)/(n+0.4)$, and calculate each failed test sample's accumulated failure probability $F(t_i)$;

(3) According to test data t_i and $F(t_i)$ of each failed test sample, draw points in the $t\text{-}F(t)$ coordinate system on Weibull probability paper;

(4) Determine the regression line according to the points drawn in Weibull probability paper, and if the points drawn are not on the same line these points should be "linearized" by the following methods:

1）Prolong the curve obtained by the points drawn in the $t\text{-}F(t)$ coordinate system on Weibull probability paper according to $[t_i, F(t_i)]$, and let it intersect with t scale at one point, whose reading is the estimate value $\hat{v}$ of position parameter v;

2）Figure out $t_i^{'}$ according to $t_i^{'} = \mathbf{v}_i - \hat{\ }$, and then draw points in the $t\text{-}F(t)$ coordinate system on Weibull probability paper according to $[t_i^{'}, F(t_i)]$, to determine the regression line;

(5) Determine the corresponding reliabiity R (T_e) on the Weibull probability paper according to the rated life T_e;

(6) Determin the corresponding product reliability grade in Table 4-2 according to R (T_e).

6 RELIABILITY TESTING DEVICE

The reliability test of relays generally adopts microcomputer for control and detection.

6.1 TECHNICAL PERFORMANCE OF TEST DEVICE

For the reliability testing of control relay, its total test time is very long. Therefore, multiple test samples and contacts should be tested, and the test device should come with multiplexed output and input. During the test, the contact voltage drop at all closed contacts and the voltage between all open contacts can be monitored at each action of the test sample, to inspect whether the contact has excessive contact voltage drop, or bridging, splicing, low insulation resistance and other faults occurring between the contacts. In addition, taking into consideration the assessment methods specified in some standards, at each action of the test sample, the test device can also assess whether the release time of its pull-in time exceeds the specified value. The contact voltage drop, voltage between open contacts, pull-in time and release time should be successively monitored. Moreover, the test device should be capable for automatic and periodic measurement of the pull-in voltage and release voltage of electromagnetic relay, each measuring interval can be adjusted randomly, and the specific values of contact resistance can also be measured periodically.

After the test, detect the test sample's insulation resistance and medium pressure-proof, etc. which can be tested one-off by general test equipment. At failure, the device can record the failed test sample no., the failure occurring time, and failure mode, sort out the data output, and alarm. The test initial parameters, such as the threshold voltage to determine excessive contact voltage drop, the threshold value of pull-in time, and total test times, can be entered into the mainframe by keyboard, and can be modified at any time by the keyboard throughout the test process.

For more flexible testing, in case of failure during the test, the device can judge whether to stop the test according to the entered control parameters. The other functions of test device include setting of test operating frequency, and nonvolatile memory. At power recovery, the device can be actuated both in automatic and manual modes, during which, the created data shall not be damaged, and the finished test times can be calculated further.

For the electromagnetic system coil of the test sample, the test device is capable of actuating both *DC* and *AC* coils, simply under the specified maximum voltage value of the actuated coil (generally up to 1000V), and only if the test sample coil voltage applied by the user does not exceed this value, the test device is capable for the operation of coil power-on and off, without any limit to the test sample contact types (normal-on, normal-off, or switching, etc.).

6.2 HARDWARE DESIGN OF TEST DEVICE

In addition to high reliability, the test device should come with some capacity for analysis and processing, to easily complete various tests, as well as the type judgment, printing and alarming of relay failure. The reliability of the microprocessor is very high, so the selection of peripheral equipment is extremely important. To further improve the reliability of the device as a whole, mature module line can be adopted. In addition, the device price should not be consequently too high, and the operating capacity of the mainframe is not necessarily emphasized during selection, but priority should be given to the control-oriented industrial control computers. For this end, the most perfect one is the industrial control computer with STD-BUS module.

The full name of STD-BUS is industrial control standard bus, and the various computer modules designed and manufactured by this standard is called STD-BUS module. The principle block diagram of the device is shown in Figure 4-2. And the principle of hardware components is briefed below.

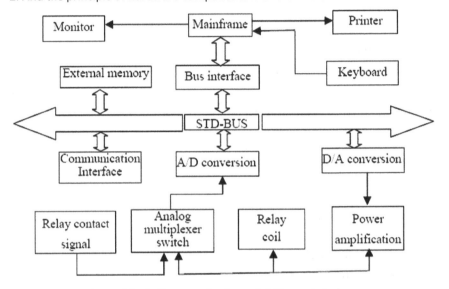

Figure 4-2 Hardware block diagram of relay reliability test device

6.2.1 MAINFRAME

The mainframe adopts microprocessor system with relatively perfect functions. Strictly, the mainframe is not STD-BUS module, but microcomputer system with STD-BUS interface, thus the complete functions and flexible adapting of the mainframe can be kept. The reliability test can be completed by the various modules that work in coordination according to specified test method, utilizing the program in the mainframe memory (the program is firmed in the mainframe memory). The mainframe directly controls the monitor, printer and keyboard, with timer hardware inside. Therefore, all data resulted from tests and test progression can be displayed, the operator can control the test parameters and arrange tests by keyboard. By the aid of keyboard, some special tests can also be conducted, and the test data can be directly output, for data analysis by advanced computer, or printing output after a little analysis and processing for analysis by the operator. The mainframe generally does not conduct complex mathematic operation, such as numerical statistics, etc.

6.2.2 D/A CONVERTER PLATE

The D/A converter plate converts the digital quantity sent from the mainframe to analog quantity, which, after power amplification, is transformed to the analog voltage that is enough to actuate multiple test samples (relays), and this voltage is generally the rated coil voltage that can complete coil power-on. When the D/A converter output voltage is zero, the coil is powered off. When the measurement of test sample pull-in voltage is required, what sent from the mainframe to the D/A converter plate is continuously increasing digital quantity, which meets the requirement of voltage boost test method, while the analog voltage applied on the coil becomes the voltage that complies with the specified voltage waveform.

In addition, the mainframe also continuously measures the voltage at both coil ends and detects the closing situation of corresponding contact at each voltage boost, to measure the test sample pull-in voltage. When the measurement of test sample release voltage is required, what sent by the mainframe is a series of continuously decreasing digital quantity, but not increasing digital quantity. Other processes are similar to the measurement of pull-in voltage. It should be noticed that the above process of voltage boost and drop is only applicable to DC coil, and the AC coil cannot be actuated by the analog voltage after simple amplification, but by other measures.

6.2.3 A/D CONVERTER PLATE

A/D converter plate converts the external analog voltage to digital quantity for the mainframe to measure: the contact voltage drop of contact, the voltage between open contacts, and coil voltage. This module is shared by multiplex analog signals through a multiplex switch. The voltage values on each contact and on each relay coil are generally sent simultaneously to the input terminal of multiplex switch. Whether they should be sent to the A/D converter plate is controlled by the mainframe. The mainframe only selects a channel signal at any moment and sends it to the input terminal of the A/D converter plate though multiplex switches. Needless voltage signals are insulated at the input terminal of multiplex switch.

Thus, the voltage at multiple pairs of contact and the voltage on multiple test sample coils can be converted by time-sharing. In addition, the pair number of tested contacts should be within 40~120. Too many or few contact pairs will be tested by batches, which will influence the testing speed. The number of test sample is generally determined by the number of contacts. However, the number of test sample scarcely influences the testing time, because the coil voltage is measured periodically instead of at each action.

6.2.4 EXTENDED EXTERNAL MEMORY

The extended external memory is used to store large amount of test data. Because the access to the external memory will occupy the mainframe for a period of time, not every test data are immediately sent to the external memory, they are first stored in the mainframe memory, and then transferred to the external memory by batch when the external memory has certain capacity. For the reliability test of hundreds of thousands of times, the internal memory capacity is extremely sufficient in general. The whole module system is installed on a special socket that can be added by modules randomly to increase the system functions.

6.3 SOFTWARE DESIGN OF TEST DEVICE

The program flowchart is shown in Figure 4-3.

6.3.1 SYSTEM INITIALIZATION

(1) System self-check: check whether the memory, printer and modules work normally.
(2) Test sample wiring check: because one test sample may come with normal open, normal close and switching contacts, the device can automatically check out the number of contact pairs and contact type on any test sample. The check result can be printed as table output, for the operator to check if the wiring is wrong.

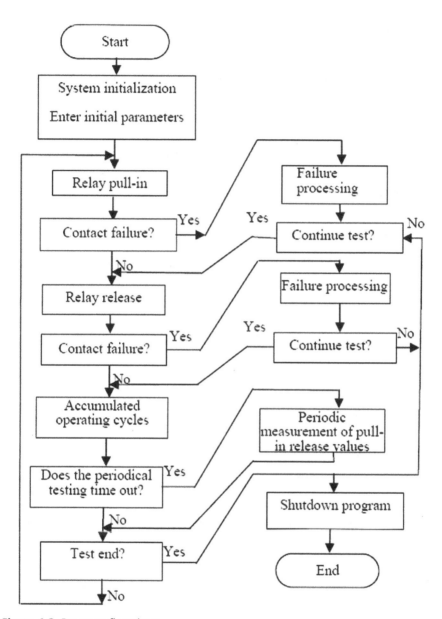

Figure 4-3 Program flowchart

The software mainly has the following program segments:

6.3.2 ENTER INITIAL PARAMETERS

(1) The operator enters the initial data to the system memory by keyboard, such as the operating frequency, total testing times, threshold value of contact voltage drop, etc.

(2) Output and printing of all initial parameters for the operator to check up. The illogical entry (usually caused by missing operation) can be screened by the program software, to complete the task of error correction.

(3) Completion of setting of various parts in the test system according to the initial parameters

6.3.3 COIL POWER UP

The mainframe applies the rated voltage on the coil through the actuator. Taking the test requirements into consideration, after a test sample lost its significance for further testing (such as the occurrence of unrecoverable fault), the mainframe will remove the test sample through a mask subprogram, when the rated voltage at coil power up will not be applied on the coil of this test sample.

6.3.4 COIL POWER-OFF

Disconnect the coil power supply.

6.3.5 JUDGMENT OF TEST SAMPLE FAILURE

(1) Judgment on excessive contact voltage drop on all contacts to be closed

(2) Judgment on low voltage between all contacts to be opened

(3) Judgment on whether the pull-in time and release time of each test sample exceeds the specified value.

6.3.6 FAULT PROCESSING

(1) In the case failure is judged, the program can analyze the failure mode, output the failure occurrence time, the failed contact No. and failed test sample No., and store the failure data into the memory for reference.

(2) According to the input initial parameters, determine whether the failed test sample should be further tested or the whole test should be ended.

6.3.7   PERIODIC MEASUREMENT

After each action is accumulated to the predetermined time, the program will complete the measurement of pull-in voltage, release voltage, voltage drop at contact and other parameters, and output the results.

6.3.8   END PROGRAM

(1) Repeat the above Periodic measurements.
(2) Sort out all data, and print detailed test report.
(3) Send shutdown signal and reset relevant modules.

6.3.9   SHUTDOWN

During the program flow, the system is in keyboard scanning state after each operation is completed, which means that the control personnel can normally pause the program flow without pausing the task in progression. If necessary, the test can be immediately stopped by the special reset key without loss of the test data results. The program is designed to protect data loss and requires a special set of complex operation instructions before the data can be cleared. This is to avoid loss of data by mistakes in operation.

5

RELIABILITY OF CONTACTOR

1 RELIABILITY INDEX

Contactor is usually used to control various motors, and it's the electrical apparatus widely applied in various control systems and equipments. The level of contactor reliability directly influences the reliability of various control systems and equipments. The reliability index of contactor can be expressed by failure rate grade, etc, as shown in Table 5-1.

Table 5-1 Failure rate grade name, symbol and maximum failure rate of contactor

Name of failure rate grade	Symbol of failure rate grade	Maximum failure rate λ_{max}(1/10 times)
Subclass V	YW	3×10^{-5}
Class V	W	10^{-5}
Subclass VI	YL	3×10^{-6}
Class VI	L	10^{-6}
Subclass VII	YQ	3×10^{-7}
Class VII	Q	10^{-7}

2 TEST REQUIREMENTS

2.1 CONDITIONS OF ENVIRONMENT

The test should be conducted in the normal application conditions specified in IEC60947-1(the temperature is (-5 to +40)°C and relative humidity is 50% to 90%)

or conducted in the environmental conditions specified in the standard or technical specification of the tested product.

2.2 CONDITIONS OF INSTALLATION

(1)   The test sample should be installed in the place of normal application

(2)   The test sample should be installed in the place without remarkable impact and vibration

(3)   The inclination between the installing surface and vertical surface of test sample should comply with the product standard

(4)   For the contactors to be installed by installation rails, the installation rails should comply with relevant standard

2.3 CONDITIONS OF TEST POWER SUPPLY

(1)   AC power supply: sine-wave power

1) Waveform distortion factor should not exceed 5%

2) The frequency is 50Hz or 60Hz with allowable deviation of ±5%

(2)   The ripple coefficient of DC power supply should not exceed 5%

2.4 LOAD CONDITIONS

(1)   To detect whether the main contacts and auxiliary contacts work normally, they can be connected to each respective detection loop, i.e. the main contact loop and auxiliary contact loop

(2)   The load can be resistance load

(3) The power supply of the contact loop should be DC 24V (or 12V) and the corresponding circuit current should be 1A (or 0.1A)

(4) When the contact is connected to the load during the test, the voltage fluctuation of contact loop should not exceed 5% to the no-load voltage

2.5 EXCITATION CONDITIONS

(1)   In the test, the sample should be energized by the rated value

(2)   The number of cycles per hour in the test should exceed the rated value specified in the product standard; in order to shorten test times, under the condition that normal working and failure mechanism of the test samples are not

affected, the number of cycles per hour can be increased to maximum 7200 times

(3) The load factor should be chosen from the product standard, or the following recommended values: 15%, 25%, 40% and 60%

3 TEST METHOD

3.1 PREPARATION OF TEST SAMPLES

Test samples should be selected randomly from the qualified products produced by lot production under the stable process conditions, and the number of products for sampling should exceed 10 times of test samples n.

3.2 INSPECTION OF TEST SAMPLES

3.2.1 INSPECTION BEFORE THE TEST

We should check whether the parts or components of test samples are damaged or broken due to transportation, eliminate the test samples with damaged parts or components, and make up the deficiency according to specified requirements. The eliminated test samples should not be counted in the relevant failure number r.

3.2.2 INSPECTION DURING THE TEST

In the test, the detection circuit should detect whether all normal open and normal closed auxiliary contacts and one main contact of the product work normally, and detect the contact voltage drop at the closed contact and the voltage between the open contacts during 40% of "on" time and 40% of "off" time of the test samples at each cycle. The test samples should not be sorted out or adjusted during the test.

3.2.3 INSPECTION AFTER THE TEST

(1) Parts or components damaged or broken
(2) Operating voltage
(3) Release voltage

3.3 FAILURE CRITERIA

In one of the following cases, the test samples are considered to be failure.

(1) Contact voltage drop U_j at closed contacts exceeds 10% of contact loop power voltage, i.e. 2.4V (or 1.2V)

(2) The voltage U_f between open contacts is lower than 90% of contact loop power voltage, i.e. 21.6V (or 10.8V)

(3) Contactor does not close when coil is energized

(4) Contactor does not open after coil is de-energized

(5) The test sample parts are seriously damaged and loosen

(6) Blocking and sticking of mechanical motion

(7) Obvious noise (caused by the broken short-circuit ring and iron core, etc.)

4 RELIABILITY GRADE

The contactor failure rate grade is determined by reliability verification test.

4.1 SAMPLING PLAN OF RELIABILITY VERIFICATION TEST

The reliability verification test of the contactor is also called failure rate test which can include: classification test, maintenance test, and upgrade test. In classification test and upgrade test, the confidence level is 0.9, and the sampling plan is shown in Table 5-2. In maintenance test, the confidence level is 0.6, the sampling plan is shown in Table 5-3.

Table 5-2 Sampling Plan for Classification Test and Upgrade Test

Failure rate grade	Test truncation times $T_c(10^6$ times)									
	A_c=0	A_c=1	A_c=2	A_c=3	A_c=4	A_c=5	A_c=6	A_c=7	A_c=8	A_c=9
YW	0.768	1.30	1.77	2.23	2.66	3.09	3.51	3.92	4.33	4.74
W	2.30	3.89	5.32	6.68	7.99	9.27	10.53	11.77	13.0	14.21
Y L	7.68	13	17.7	22.3	26.6	30.9	35.1	39.2	43.3	47.4
L	23.0	38.9	53.2	66.8	79.9	92.7	105.3	117.7	130.0	142.1
Y Q	76.8	130	177	223	266	309	351	392	433	474
Q	230	389	532	668	799	927	1053	1177	1300	1421

Table 5-3 Sampling Plan for Maintenance Test

Failure rate grade	Max. maintenance cycle (month)	Test truncation times $T_c(10^6$ times)									
		$A_c=0$	$A_c=1$	$A_c=2$	$A_c=3$	$A_c=4$	$A_c=5$	$A_c=6$	$A_c=7$	$A_c=8$	$A_c=9$
YW	24	0.306	0.673	1.03	1.39	1.75	2.10	2.45	2.80	3.15	3.50
W	24	0.916	2.02	3.10	4.18	5.25	6.30	7.35	8.40	9.44	10.5
Y L	24	3.06	6.73	10.3	13.9	17.5	21	24.5	28	31.5	35
L	48	9.16	20.2	31.0	41.8	52.5	63.0	73.5	84.0	94.4	105
Y Q	48	30.6	67.3	103	139	175	210	245	280	315	350
Q	48	91.6	202	310	418	525	630	735	840	944	1050

4.2 TEST PROCEDURE OF CONTACTOR RELIABILITY VERIFICATION TEST

4.2.1 CLASSIFICATION TEST

Procedure for classification test:

(1) To select the failure rate grade, grade YW or W should be generally selected for the first test.

(2) To select the allowable number of failure of A_c and truncation failure number $r_c(r_c=A_c+1)$, A_c is recommended within 2~5.

(3) According to the selected failure rate grade and A_c, find the truncation times T_c from Table 5-2.

(4) To select test sample deadline t_z, t_z is recommended not to be lower than 10^5 times.

(5) According to T_c, A_c and t_z, determine the number of test samples n by Formula (5-1), i.e.

$$n = \frac{T_c}{t_z} + A_c \qquad (5\text{-}1)$$

It shall be noticed that the number of test samples n is generally no less than 10.

(6) To select n samples at random from the qualified products which are produced in the same lot.

(7) To conduct test and detection complying with the test method given in this chapter.

(8) To make statistics of relevant number of failures r and the relevant test times (failure occurrence time) of each failed test sample; for the relevant failed test sample after test, the relevant test times should be calculated according to the test deadline.

(9) To make statistics of the accumulative relevant test times T.

(10) Judgment of test results: when the relevant failure number r does not reach the truncation failure number r_c (i.e. $r{\leq}A_c$), while the accumulative relevant test times T reaches or exceeds the truncation times T_c, it will be judged as qualified test(accept); when the accumulative relevant test times T does not reach or exceeds the truncation times T_c, while the relevant failure number r reaches the truncation failure number r_c ($r{>}A_c$), it will be judged as unqualified test(rejection).

4.2.2 MAINTENANCE TEST

In general, for the qualified products of classification test, the maintenance test should be conducted according to the maintenance cycle given in Table 5-3 and by the following procedures:

(1) To select the allowed failure operations A_c

(2) To inquire the truncation times T_c from Table 5-3 according to the determined failure rate grade and the determined allowable failure number of the product.

(3) To determine test sample deadline t_z (the method is same as the classification test).

(4) To determine the test sample number n (the method is same as the classification test).

(5) To select the test samples (the method is same as the classification test).

(6) To conduct test and detection complying with the test method given in this chapter.

(7) To make statistics of relevant number of failure and the relevant test times(the method is same as the classification test).

(8) To make statistics of the accumulative relevant test times T.

(9) To judge the test result (the method is same as the classification test).

(10) To conduct further the next maintenance test according to the specified maintenance cycle if the maintenance test is qualified; to re-conduct the

classification test to determine its failure rate grade if the maintenance test is unqualified.

(11) When re-determining the failure rate grade, we should accumulate all product test data from the primary classification test (including the unqualified maintenance test data), and determine the product failure rate grade according to Table 5-2 in order to determine the accumulative relevant failure number and accumulative relevant test times.

4.2.3 UPGRADE TEST

Upgrade test can be further conducted on the qualified products after classification test. The upgrade test data can be obtained by conducting the prolonged test on the sample of classification test and maintenance test, and by conducting the test on the sample for the upgrade test. The upgrade test should be conducted by the following procedures:

(1) Select the failure rate grade to be upgraded (which is generally one class higher than the originally determined one).

(2) Select the allowable failure number A_c.

(3) Inquire the truncation times T_c from Table 5-2 according to the determined failure rate grade and the determined allowable failure number of the product.

(4) Determine the prolonged test times, the test sample number and test times for the upgrade test according to T_c.

(5) Select the test sample (the method is same as the classification test).

(6) Conduct test and detection complying with specified.

(7) Make statistics of relevant number and accumulative relevant test times T.

(8) Judge the test result (the method is same as the classification test).

(9) Conduct the maintenance test of the same class according to the specified maintenance cycle if the upgrade test is qualified; to re-conduct the classification test to determine its failure rate grade if the upgrade test is unqualified.

(10) When re-determining the failure rate grade, accumulate all test data of the product, and determine the product failure rate grade according to Table 5-2 for the accumulative relevant failure number and accumulative relevant test times.

5 DETERMINED METHOD OF CONTACTOR RELIABILITY IN PRACTICAL LOAD CONDITION

5.1 LOAD OF MAIN CONTACT IN PRACTICAL USE

Apply rated voltage, rated current and resistance load as recommended.

5.2 TYPE OF RELIABILITY TEST

Reliability determination test can be adopted.

5.3 PROGRAMM OF RELIABILITY DETERMINATION TEST

(1) The number of the test samples is n, which generally can be within the range of 10 to 20.

(2) n samples are selected randomly from the qualified products from lot production, complying with the regulations given in 4.2.1 of this chapter.

(3) The samples are tested and inspected according to the method given in this chapter, and judge whether the test sample has failed to comply with the failure criteria in section 3.3 in this chapter.

(4) To stop the test when more than 50% of the samples failed.

5.4 JUDGMENT OF RELIABILITY TEST RESULT

(1) To make statistics of the relevant test times of failed samples, the relevant test times should be calculated according to the times as the test deadline. Failure occurrence time t_i is for the relevant failed samples with inspected after test.

(2) When $n>20$, $F(t_i)=i/n$; when $n \leq 20$, $F(t_i)=(i-0.3)/(n+0.4)$, and to calculate the cumulative failure probability $F(t_i)$ of each failed sample.

(3) According to the test data t_i and $F(t_i)$ of each failed sample, draw points in the t-$F(t)$ coordinate system on Weibull probability paper.

(4) To determine the regression line according to the points drawn in Weibull probability paper, if the drawn points are not on a line, they should be "linearized" by the following method.

　　1）To prolong the curve obtained by the points drawn in the t-$F(t)$ coordinate system on Weibull probability paper according to $[t_i, F(t_i)]$, and make it intersect with t scale at one point, whose reading is the estimate value $\hat{v}$ of position parameter v.

2）To figure out t_i' according to $t_i'=t_i-v$, then draw points in the t_i'-$F(t)$ coordinate system on Weibull probability paper according to [t_i', $F(t_i)$], to determine the regression line.

(5) To determine the reliable life t_R of the reliability R on the Weibull probability paper according to the determined reliability R (generally determined as $R=0.9$).

(6) To determine the product reliability under applied application load condition according to t_R and R.

6 RELIABILITY TEST DEVICE

6.1 TECHNICAL PERFORMANCE OF TEST DEVICE

The test device is capable for controlling 32 AC contactors simultaneously for reliability test, complying with the sampling requirement of checked reliability index. We can monitor whether each test sample is operating normally or not. The test sample fault can be inspected automatically, and operating times at the fault occurrence, the fault type, and the fault sample number can be printed.

When the fault number of a test sample is accumulated n (n is entered by keyboard), it can be eliminated automatically, to prevent power supply fault caused by coil burning due to test sample armature blocking.

The load factors are different from various contactors at life test, so the operating frequency, load factor and other parameters can be set and adjusted independently by the user in use of test device.

For each test sample, a pair of normal open contacts signals can be detected (or several pairs of normal open contacts can be connected in series), so there are 32 detection signals in total. When the test sample operated normally, the inspected signal should be logic "1"; when the test sample does not operate, the detected signal should be logic "0".

6.2 HARDWARE DESIGN OF TEST DEVICE

The hardware block diagram of the device is shown as Figure 5-1. Except for the input and output circuits which is operating and inspecting the test sample, all other circuits are complied with the typical structure of general peripheral circuit.

6.2.1 INPUT PART

The input part is composed of voltage divider and input buffer, etc.

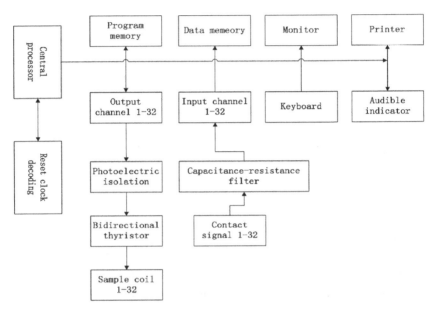

Figure 5-1 Hardware block diagram of contactor reliability test device

The voltage (5V) at input end is divided from the DC voltage (220V). The dust membrane should be adhering to the contact when contactor operates. If the voltage (5V) is directly applied at both contact ends, it usually makes misjudgment by not penetrating the membrane. When the voltage divider is adopted, the membrane can be generally penetrated by the electric field generated by the voltage (220V) on both ends. To eliminate the pulse interference on the input end by adopting capacitance-resistance filter. The contact closing situation is fetched by CPU through the buffer and bus actuator for processing.

6.2.2 OUTPUT PART

The output part is composed of latch, amplifier circuit, photoelectric isolation circuit and bidirectional thyristor circuit. After setting data to the latch by CPU, the corresponding optical coupler can be controlled on/off by the data. The trigger end of bidirectional thyristor is added with the corresponding trigger signal to turn on the thyristor or turn it off after zero current.

6.2.3 INTERFERENCE RESISTANT MEASURE

In actual application, the test equipment is usually located in severe environment. The frequent actuating of various high power electrical machinery (such as motor, welding machine and crane) may lead to instantaneous grid voltage drop, or spike pulse interference or strong electrical noise interference with wide spectral range. Therefore, the hardware shall be designed to come with corresponding interference resistant measures for the input and output parts.

First the output part is equipped with photoelectric isolating device, to realize electrical isolation of the microcomputer part and the control output part, to prevent ground loop. The input part is equipped with capacitance-resistance filter, to eliminate the pulse interference at the input end effectively.

The power input end is equipped with 220V AC stabilizer to reduce the influence of grid fluctuation. Then it is connected to the regulated power supply dedicated for microcomputer through a low-pass filter for microcomputer power supply.

In terms of wiring, the heavy current and weak current are separated as possible as they can be. The input lead and output lead are complied with twisted pair to eliminate electromagnetic interference. The bidirectional thyristor is installed in the printed circuit board independently to avoid harmful effects to host from strong radiation.

6.3 SOFTWARE DESIGN FOR TEST DEVICE

6.3.1 DESIGN PRINCIPLE

To reduce the impact current of power supply, at program design, the sample driver is designed as itinerant driver, and the contact is detected at corresponding time to judge whether work normally. First the 32 test samples are divided into four groups, which are driven and detected according to specified sequence, until the detection is completed.

To increase flexibility, the software adopts the structure of program modules. It mainly includes:

(1) Initiation module segment

(2) Operating/release module segment

(3) Fault estimation module segment

(4) System service program module segment

To increase the reliability of software running, the program is added with self-correcting software when running.

6.3.2 SOFTWARE FUNCTIONS AND PROGRAMMING

(1) Initialization module is to complete the system reset operation, and receive the various control parameters entered by the operator through keyboard, such as operating frequency, etc.

(2) Operating/release module is to complete the operation and release of a test sample.

(3) Main program module, as a core module, gives the corresponding control signal and deploys the operation/release subprogram, to realize test sample cycle operation. The block program of main program module is shown as Figure 5-2.

(4) Fault estimation module: because there are 32 input signals in total with time difference, the fault judgments are not synchronous, but controlled at the moment when each test sample should be detected to avoid bounce time of the contact. If any unreliable close or open during the detection, the software will record the fault time, fault sample number, fault type, and the accumulative fault times of the test sample automatically.

(5) System service program module mainly comes with the functions of controlling the printer, converting various digital mode systems, and a series of system services, such as display and keyboard.

6.3.3 SOFTWARE RUNNING MODE

The software runs in three modes:

(1) Continuous testing after record and printing in case of fault: this mode is mainly applied in the case without the operator.

(2) Automatic delay after recording and printing in case of fault, waiting for the observation and handling by personnel, the delay time is entered and set by the test personnel, several minutes in general.

(3) Auto standby state after recording and printing in case of fault, until restarted by operator.

After the test is completed, it is capable for one-off auto testing the action time of each test sample and printing output, and then sorting out all test data during the test process and output the detailed test report on the printer. Since the data memory of the equipment comes with accumulator protecting circuit, the data can be canceled only after a series of complex operations. The data will not be lost in case of power off and reset, etc. The running test can recover automatically after power recovery, with completely successive parameter data of both tests.

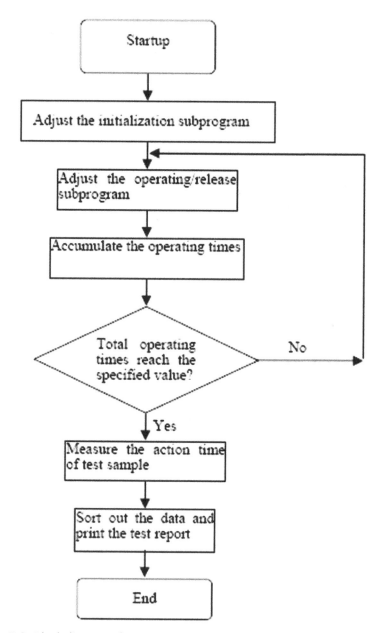

Figure 5-2 Block diagram of main program

RELIABILITY OF MINIATURE CIRCUIT BREAKER

1   RELIABILITY INDEX

Miniature circuit breaker (MCB) is a low-capacity molded case circuit breaker with a very compact structure. It branches out from the low-voltage circuit breaker and now has become one category of terminal electrical apparatus. According to the definition of IEC898 standard and the Chinese standard GB10963 *Circuit Breakers for Household and Similar Installations,* the miniature circuit breaker refers to "the AC circuit breakers with AC 50Hz, voltage 380V and below, current 125A and below, and the rated limiting short-circuit breaking capacity of not more than 25000A".

With the improvement of electrification degree in daily life and the development of urban construction, the consumption of non-industrial electricity has been increasing, so the safe operation and protection of electrical equipment has become an important issue. In this case, the traditional approach used by a combination of the knife switch and fuse can not meet the requirements and thus the demand for high - performance MCB has been increasing rapidly, and the annual output of MCB is also rising continually. The reliability level of MCB can directly affect the safe operation of facilities on distribution lines, it is closely linked to the daily life of millions of households; meanwhile, due to its large output and wide application, the failure of MCB may bring severe economic losses to its users if its reliability level is not high. Therefore, it is urgent and necessary to study the reliability of MCB, which not only has become a significant work in the field of electrical apparatus but also is profound to guarantee and improve the quality and reliability of MCB.

With the continuous development of the MCB technology, the product technical specifications and standards have been formulated successively by standard commissions of different countries. The 'Circuit Breakers for household and Similar Installations' Sub-Committee of International Electrotechnical Commission (SC23E) formulated the standard IEC898 – 1987 *Circuit Breakers for home application and Overcurrent Protection Circuit Breaker for Similar Installations* in 1987. In recent years, China has actively accepted or equivalently adopted the guidelines of IEC international common standards in terms of electrical apparatus. In order to promote the development of MCB, the Chinese standard GB10963-1989 *Circuit Breakers for Household and Similar Installations*, was officially issued and implemented in 1989 following the publication of the IEC898-1987 standard. In 1995, IEC amended the above standard into the IEC60898: 1995 and in 2005, China also adopted IEC standards to formulate the GB10963-2005 standard. Hebei University of Technology, Shanghai Electrical Apparatus Research Institute (Group) Co., Ltd. and other institutions formulated the Chinese industry standard JB / T 10493-2005 *Reliability Test Method of Overcurrent Protection Circuit-breakers for Household and Similar Installation.*

1.1 OPERATING FEATURES AND FAILURE MODES OF MCB

MCB belongs to the protective electrical apparatus used to protect the safety of the distribution lines and equipments, it is different from the control electrical apparatus which are operated frequently such as relays used in control circuits, small-capacity AC contactors and so on. When the overload current and short-circuit occurs on the low voltage distribution lines or electrical equipments, the release should be able to trip in time and cut off the circuit reliably; when the distribution lines or electrical equipments are in normal operating state, its main contacts can connect the circuit reliably and the release should not trip. In addition, MCB is one of the protective electrical apparatus without frequent operations. Its electrical life specified in the product standard is far lower than that of the control relays and the contactors, which is merely about thousands of times in general.

The main failure modes of MCB are divided into the following three categories:

(1) Operation failure, i.e., MCB cannot be switched on when performing manual closing operation and the circuit cannot be connected; MCB cannot be switched off when performing manual switch-off operation and the circuit cannot be disconnected.

(2) Wrong action failure, i.e., when the overload or short circuit current doesn't occur on distribution lines or electrical apparatus, the instantaneous release or the delayed-action release of MCB automatically cuts off the circuit.

(3) Refusing action failure, i.e., when the overload or short circuit current occurs on distribution lines or electrical apparatus, the MCB cannot cut off the fault current in time, and then not guarantee the reliable protection of the distribution lines or electrical apparatus.

In general, overload or short-circuit current does not occur frequently on lighting circuits or electrical equipment, so some circuit breakers may not endure overload or short circuit fault in several years while some may endure several faults in one year. Therefore, the refusing action failure does not occur frequently in general. In addition, in consideration of the consequence of failure, although operation failure or wrong action failure may cause abnormal electricity connection or unnecessary power outages from lighting circuits or electrical equipment, consequently result in certain economic losses, but usually does not cause severe consequences. However, the refusing action failure may endanger the safety of the lighting circuits or electrical equipment, even cause fire outbreak in buildings which can lead to greater economic losses, the consequence of refusing action failure is much more severe than the other two kinds of failure.

1.2 RELIABILITY INDEX SYSTEM

It is known from the operating features and the failure modes of MCB that it is difficult to describe its reliability level based on individual reliability index, so the Chinese industrial standard JB/T 10493-2005 has stipulated two reliability indices on refusing action failure, wrong action failure and operation failure. For the former two kinds of failure, the value of protection success ratio R could be used as one of reliability indices (Success ratio refers to the probability of products to perform the required functions under stated conditions for a specified period of time, or the probability of test success under stated conditions); for the latter kind of failure, the value of operation failure rate λ could be used as another reliability index. (Operation failure rate refers to the probability of failures occurred in unit time after the moment t when the product has operated until the moment t)

The MCB protection of electrical equipments includes overload protection and short circuit protection. As damages caused by short circuit current on electrical equipments is much more severe than that caused by overload current, and in order to avoid the reliability index system to be too complex, it is unsuitable to use too much reliability indices. So, it could be mainly considered the short circuit protection reliability of MCB for lines or equipments. Therefore, the instantaneous protection success ratio R (hereinafter referred to as the success ratio) and the operation failure rate λ (hereinafter referred to as the failure rate) have been adopted as the MCB reliability characteristics and the grades of failure rate and success ratio have been set as the MCB reliability indices separately.

The value of the maximum failure rate λ_{max} is recommended to be divided into three failure grades (Grade III, Subgrade IV, and Grade IV) and the value of the unacceptable success ratio R_1 can be divided into five grades (Grade I, Grade II, Grade III, Grade IV, Grade V). Table 6-1 shows the failure rate grade and the maximum failure rate λ_{max}. Table 6-2 shows the success ratio grade and the unacceptable success ratio R_1.

Table 6-1 Name of MCB failure rate grade and the maximum failure rate λ_{max}

Name of failure rate grade	Maximum failure rate λ_{max}（1/10times）
Grade III	1×10^{-3}
Subgrade IV	3×10^{-4}
Grade IV	1×10^{-4}

Table 6-2 Name of MCB success ratio grade and the unacceptable success ratio R_1

Name of success ratio grade	R_1
Grade I	0.99
Grade II	0.98
Grade III	0.97
Grade IV	0.96
Grade V	0.95

2   TEST REQUIREMENTS

2.1 TEST SITE

The reliability test could be either the laboratory test or the field test. Their advantages and disadvantages are shown in Table 6-3.

The MCB reliability verification test is recommended to employ the laboratory test.

2.2 TEST CONDITIONS

2.2.1   ENVIRONMENTAL CONDITIONS

(1) Temperature: 15 °C to 35 °C; relative humidity: 50% to 90%; atmospheric pressure: 86 to 106kPa.
(2) Avoiding dust and other pollutants.
(3) The test sample should be placed in the standard atmospheric conditions for enough time (not less than 8h), in order to make the test sample reach the thermal equilibrium.

Table 6-3 Contrast of advantages and disadvantages of the laboratory test and the field test

Test	Field Test	Laboratory Test
Advantage	1.Need less test equipments	1. Able to limit and control the test conditions
	2.Can provide more practical test results	2.Able to control the supervision and the failure display of samples' performance in a better condition
	3.Lower test cost	3.Reproducibility and comparability of test results
Disadvantage	Unable to proceed under rigorous control conditions	

2.2.2 INSTALLATION CONDITIONS

(1) The test sample should be installed according to the stipulations of the product standard.
(2) The test sample should be installed in the place without remarkable impact and vibration.
(3) The inclination between the installing surface and the vertical surface of the test sample should be consistent with the requirements of product standard.
(4) Miniature circuit breakers that installed by installation rails should be in accordance with the installation rail standards

2.2.3 CONDITIONS OF TEST POWER SUPPLY

(1) AC power supply should be 50 Hz sine wave power with an allowable deviation ±5%
(2) DC power supply may adopt generator, accumulator or stabilized voltage power supply.
(3) When the contact is connected to the load during the test, the voltage fluctuation of test power supply relative to the no-load voltage should not exceed 5%.

2.2.4 LOAD CONDITIONS

(1) Load power supply can be DC or AC power supply, DC power supply is recommended in general.

(2) The load can be resistive load, inductive load, capacitive load, and non-linear load, resistive load is recommended in general.

(3) Power voltage U_e in contact loop should be 24V/6V or the minimum DC voltage rating specified in the product standards

(4) Load current in contact loop I_c should be 1A/100mA.

2.2.5 EXCITATION CONDITIONS

(1) Operational reliability test: For the circuit breakers with $I_e \leq 32A$, the recommended operating frequency is 240 cycles/h, for each cycle, the recommended time interval on off-position is more than 13s; for the circuit breakers with $I_e > 32A$, the recommended operating frequency is 120 cycles/h, for each cycle, the recommended time interval on off-position is more than 28s.

(2) Instantaneous protection reliability test: For C-type circuit breakers, energizing current is $10\,I_e$ and $5\,I_e$; for D-type circuit breakers, energizing current is $50\,I_e$ and $10\,I_e$.

3 TEST METHOD

3.1 PREPARATION OF TEST SAMPLES

The test samples used in tests should be randomly selected from the qualified products produced by batch production under the stable process conditions.

3.2 INSPECTION OF TEST SAMPLES

Inspection items of MCB reliability test are shown in Table 6-4.

3.3 FAILURE CRITERIA

When any of the following cases occurs during the operational reliability test, it is deemed to be one failure.

(1) Contact voltage drop U_j across the closed contacts exceeds the following limit value U_{jx}. The limit value of contact voltage drop U_{jx} is 5% or 10% of the rated voltage in contact circuit.

(2) The voltage across the open contacts U_c is lower than the limit value U_{cx}. The value of U_{cx} should be 90% of the rated power voltage of contact loop unless it is accordance with the product standard stipulates.

(3) Contact welding or sticking.

(4) Miniature circuit breakers cannot close on closing operation command.

(5) Miniature circuit breakers cannot open on opening operation command.

(6) There is destructive damage in the parts or components of test samples, or loose connection between wires and parts.

After the operational reliability test, test samples should undergo the appearance inspection, insulation resistance and dielectrics strength. If any result of the above inspections doesn't meet the product standard requirement, the sample is deemed to have one failure.

When any of the following cases occurs during the instantaneous protection reliability test, it is deemed to be one failure.

(1) When the value of current I passing through C -type circuit breaker is equal to 10 times that of the rated current I_e (D-type circuit breaker is energized by 50 I_e), the break time of the circuit breaker is no less than 100ms.

(2) When the value of current I passing through C -type circuit breaker is equal to 5 times that of the rated current I_e (D-type circuit breaker is energized by 10 I_e), the break time of the circuit breaker is less than 100ms.

Table 6-4 Inspection items of MCB reliability test

Items	Operational Reliability Test	Instantaneous protection reliability test
Before Test	1. To check whether the parts or components of test samples are damaged or broken due to transportation	It is carried out before and after the operational reliability test.
	2. To eliminate the damaged test samples and replace them by new ones according to the specified requirements; the eliminated test samples shouldn't be counted into the number of failures r	
During Test	1. To detect the contact voltage drop across the closed contact in 40% "switch-on duration" of samples for each cycle	1. Starting from the cold state, C-type circuit breakers are energized 10 I_e (D-type circuit breakers are energized 50 I_e), during the test, it is to detect whether the samples can trip within 100ms
	2. To detect the open circuit voltage across the contacts in 40% "switch-off duration" of samples for each cycle	2. Back to the cold state, C-type circuit breakers are energized 5 I_e (D-type circuit breakers 10 I_e), during the test, it is to detect whether the breaking time of samples is no less than 100ms
After Test	1. Appearance inspection	
	2. Insulation resistance	—
	3. Dielectric strength	

4 SAMPLING PLAN AND TEST PROCEDURES OF RELIABILITY VERIFICATION TEST

4.1 SAMPLING PLAN OF RELIABILITY VERIFICATION TEST

The sampling plan of reliability verification test has two kinds: curtailed sequential test and time or failure curtailed test plan. Both of them have advantages and disadvantages. Curtailed sequential test needs lest average number of failures and average accumulative test time when a judgment is given, but its necessary maximum accumulative test time or number of failures may exceed those of the corresponding time curtailed or failure curtailed test. In addition, its test cost and number of samples may also change much and some difficulties exist in the arrangement of samples, test equipment, staff and so on. Therefore, the curtailed sequential test is generally applicable for the expensive products. As the time curtailed or failure curtailed test can confirm the maximum accumulative test time or the maximum number of samples before test, it is possible to determine the maximum demand of cost, test equipment and staff, so it is generally applicable for the inexpensive products.

Therefore, it is suggested to adopt the failure curtailed test for MCB failure rate verification test and MCB success ratio verification test.

4.1.1 SAMPLING PLAN OF FAILURE RATE VERIFICATION TEST

The determination method of the sampling plan of failure rate verification test is similar to that of 4.1 in Chapter 4, so the failure rate verification test plan is shown in Table 6-5.

Table 6-5 Failure rate verification test plan （β=0.1）

Truncation time T_C (10^4 times) / Allowable number of failures A_C / Failure rate grade	0	1	2	3	4	5	6	7	8
Grade III	2.3	3.89	5.32	6.68	7.99	9.27	10.53	11.77	13.0
Subgrade IV	7.68	13.0	17.7	22.3	26.6	30.9	35.1	39.2	43.3
Grade IV	23.0	38.9	53.2	66.8	79.9	92.7	105.3	117.7	130

4.1.2 SAMPLING PLAN OF SUCCESS RATIO VERIFICATION TEST

Success ratio is the reliability characteristic which is shared by all protective electrical apparatus including MCB, proposed aiming at the refusing action failure or wrong action failure in service.

According to the theory of the failure curtailed verification test for success ratio, the relation curve between the acceptance probability $L(R)$ and the product success ratio R is shown in Figure 6-1 under four parameters R_0, R_1, α and β. The following relation can be obtained from Figure 6-1.

$$L(R_0) = 1 - \alpha \tag{6-1}$$

$$L(R_1) = \beta \tag{6-2}$$

The following relation can be obtained based on derivation of (6-1) and (6-2)

$$\sum_{r=0}^{A_c} C_n^r R_0^{n-r} (1 - R_0)^r \geq 1 - \alpha \tag{6-3}$$

$$\sum_{r=0}^{A_c} C_n^r R_1^{n-r} (1 - R_1)^r \geq \beta \tag{6-4}$$

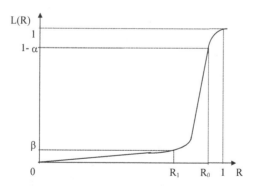

Figure 6-1 Sampling characteristic curve of success ratio

The success ratio verification test plan can be determimed through calculating the minimum integer solution from the above fomula. Table 6-6 lists some of success ratio verification test plans. In the formula

$D_R = (1 - R_1)/(1 - R_0)$ in Table 6-6, D_R is generally called as discrimination ratio; r_{RE} is the rejection number of failures, $r_{RE} = A_c + 1$.

Table 6-6 Failure curtailed verification test plan for success ratio under R_0, R_1, α and β

R_0	D_R	α=0.05 β=0.05		α=0.10, β=0.10		α=0.20, β=0.20	
		m	r_{RE}	m	r_{RE}	m	r_{RE}
0.99	1.50	5320	66	3215	40	1428	18
	1.75	2581	35	1607	22	714	10
	2.00	1567	23	945	14	453	7
	3.00	521	10	308	6	142	3
0.98	1.50	2620	65	1605	40	713	18
	1.75	1288	35	770	21	356	10
	2.00	781	23	471	14	226	7
	3.00	259	10	153	6	71	3
0.97	1.50	1720	64	1044	39	450	17
	1.75	835	34	512	21	237	10
	2.00	519	23	313	14	150	7
	3.00	158	9	101	6	47	3
0.96	1.50	1288	64	782	39	337	17
	1.75	625	34	383	21	161	9
	2.00	374	22	234	14	98	6
	3.00	117	9	76	6	35	3
0.95	1.50	1014	63	610	38	269	17
	1.75	486	33	306	21	129	9
	2.00	298	22	187	14	78	6
	3.00	93	9	60	6	28	3

The verification test plan for success ratio may also be determined based on R_1、β. Table 6-7 shows the success ratio verification test plan under R_1、β.

Table 6-7 Success ratio verification test plan （β=0.1）

m \ A_c R_1	1	2	3	4	5	6	7	8
0.95	77	105	132	158	184	209	234	258
0.96	96	132	166	198	230	262	292	323
0.97	129	176	221	265	308	349	390	431
0.98	194	265	333	398	462	525	587	648
0.99	388	531	667	798	926	1051	1175	1297

4.2 TEST PROCEDURES

4.2.1 TEST PROCEDURES OF FAILURE RATE VERIFICATION TEST

The failure rate verification test can be conducted according to the following procedures:

(1) To select the failure rate grade: Grade III or Subgrade IV is recommended for the first test.

(2) To select the allowable number of failures A_C and the truncation number of failures $r_C = A_C + 1$, A_C is recommended within 1~5.

(3) To find the truncation time T_C from Table 6-5 according to the selected failure rate grade and A_C'.

(4) To select the test deadline t_z of test samples, and t_z should not exceed the number of the electrical life specified in the product standard and is recommended to be equal to 6000 operations. ($t_z = 6000$ ops)

(5) According to T_C, A_C, t_z, the number of test samples n is determined by the formula:

$$n = \frac{T_C}{t_z} + A_C \tag{6-5}$$

(6) To select n samples randomly from the qualified products by batch production. The number of products for sampling should be no less than 10 times that of the samples n, $N \geq 10n$.

(7) To conduct the test and inspections specified in the above reliability test methods.

(8) To make statistics of the relevant number of failures r and the relevant test time of each failed samples. The relevant test time of failed samples after test is counted as its test time at the end of test.

(9) To make statistics of the relevant accumulated test time T.

(10) Judgment of test results.

The test will be judged as qualified test (acceptance) when the accumulative relevant test time T reaches or exceeds the truncation time T_C, but the relevant number of failures r is no more than the truncation number of failures r_C (i.e. $r \le A_C$) ; the test will be judged as unqualified test (rejection) when the accumulative relevant test time T is less than the truncation time T_C, but the relevant number of failures r reaches or exceeds the truncated number of failure r_C ($r > A_C$) .

4.2.2 TEST PROCEDURES FOR SUCCESS RATIO VERIFICATION TEST

The success ratio verification test will be performed according to the following procedures:

(1) To select the success ratio grade.

(2) To select the allowable number of failure A_C ;

(3) To find the number of tests n_f required for accepetance judgement from Table 6-7 according to the selected success ratio grade and A_C ;

(4) To select the truncation number of test n_z for each sample, $n_z = 10$ times in general;

(5) According to n_f , n_z and A_C , the number of samples n is determined by the formula (6-6):

$$n = \frac{n_f}{4n_z} + \frac{A_C}{2} \qquad\qquad (6\text{-}6)$$

(6) To select n samples randomly from the qualified products by batch production. The number of products for sampling should be no less than 10 times that of samples n , $N \ge 10n$.

(7) To conduct the test and inspections specified in the above reliability test methods.

(8) To make statistics of the total number of failures for all samples r_d ($r_d = r_1 + r_2$); r_1 refers to the number of refusing action failure while r_2 refers to the number of wrong action failure in the formula.

(9) Judgment of test results.

The test will be judged as qualified test (acceptance) when the accumulative number of tests n_Σ is no less than the number of tests n_f which is required for making judgement whilst the total number of failures r_d is no more than truncation number of failures r_C ($r_d \leq A_C$); the test will be judged as unqualified test (rejection) when the accumulative number of tests n_Σ is less than the number of tests n_f whilst the total number of failures r_d is more than the truncation number of failures r_C ($r_d > A_C$).

5 RELIABILITY TEST DEVICE

5.1 TECHNICAL FUNCTIONS OF TEST DEVICE

The main technical functions of the test device are as follows:
(1) Users can change and set the reliability test parameters, such as the limit value of the contact voltage drop U_{jm} of the closed contacts, the operation frequency of test samples, the total test time and so on.
(2) The test device can simultaneously perform the operational reliability test and the instantaneous protection reliability test of 8 samples
(3) The test device can simultaneously detect the contact voltage drop and the open circuit voltage of 32 pairs of contacts.
(4) During the instantaneous protection reliability test, the test device can automatically adjust the test current value in the sample loop according to test requirements.
(5) It can automatically record test data and print the data such as the test time, the serial number of the failed sample, the failure time, the serial number of the failed contacts and the failure type, as well as automatically remove the failed samples.
(6) For the operational reliability test , the operation frequency of the sample can be adjusted within 10 - 500 operations per hour and the load factor is also adjustable; for the instantaneous protection reliability test , the test current provided by the test device can be adjusted within 0 ~ 900A continuously.
(7) It can provide 4 groups of resistive load (24V or 6V and 1A or 0.1A). The external load is allowable.

(8) It is easy to operate with good man-machine interface.

5.2 HARDWARE DESIGN OF TEST DEVICE

The structural block diagram of this test device is shown in Figure 6-2. The reliability test device of MCB consists of four parts: the test control cabinet, the test sample cabinet, the large current test cabinet and the test load cabinet.

1. The test control cabinet is composed of the industrial controlling computer, the printer and the test controlling circuit boards. It is mainly responsible for completing all computer control and detection during the test.

2. The sample cabinet is equipped with 8 manipulators driven by motors to control 8 samples to switch on /switch off; meanwhile each manipulator is equipped with switch-on/switch-off signal detection circuits and brakes to ensure the accuracy of the switch- on /switch- off operation. Both of above operations are conducted under computer control.

3. The large current test cabinet consists of the large current transformer, the voltage regulator, the sampling resistor, the acquisition card, the contactor of main control loop and 8 contactors of by-pass control loops. The switch-on/switch-off operation of the contractors is also controlled by the computer. According to the test requirements, the test device adjusts the test loop current of each sample successively through adjusting the regulator to make the current to be 5 /10 times of the rated current in order to perform instantaneous protection reliability test.

4. The test load cabinet could provide four groups of resistive load and the test can be conducted under two voltages 24V or 6V and two currents 1A or 0.1A.

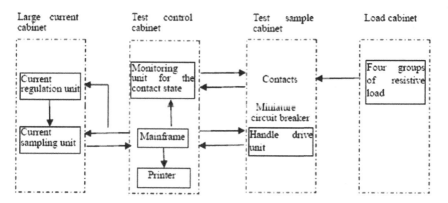

Figure 6-2 Structural block diagram of the test device

The operation principle of the main units of the test device will be described as follows.

5.2.1 MAINFRAME

The mainframe plays a crucial role in the whole test device. It is responsible for all controlling and detecting operations relevant to the test, such as the sending and receiving of test parameters; the monitoring of the test sample contact state; the forward rotation and reverse rotation of the motor and the signal acquisition which shows whether the sample handle switches on/off in place; the sampling and adjustment of test current in the instantaneous protection reliability test; the storage of test data and the analysis of failure data. Therefore, the mainframe of this test device is chosen as the industrial control computer with high anti-jamming capability based on the PC bus standard (hereinafter referred to as ICC).

5.2.2 SWITCH ON/OFF DRIVE UNIT OF THE TEST SAMPLE HANDLE

This functional unit conducts the switch-on/switch-off operation of the test sample handle depending on the computer controlling of positive inversion of the motor and stop the motor when receiving the signal that shows the handle has been switched on/off in place. Therefore, the sample can be switched on/off under the control of computer according to the selected operation frequency. Figure 6-3 shows the block diagram of this circuit.

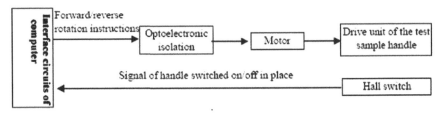

Figure 6-3 The block diagram of the handle switch-on/switch-off drive unit

It can be seen from the operation principle and the structure, this functional unit can be mainly divided into two parts: the circuit driving the sample handle to switch on/off and the circuit monitoring whether the handles are switched on/off in place.

5.2.3 MONITORING UNIT OF CONTACT STATE

The main function of this functional unit is to collect the signals of actual connected/disconnected state for all contacts of the test sample when the sample handle switches on/off in place and to send the state signals to the computer after distinguishing and comparing the condition. Figure 6-4 shows the functional block diagram of this circuit.

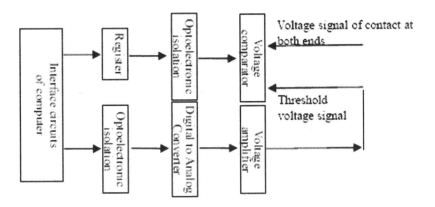

Figure 6-4 Block diagram of monitoring unit of contact state

Seen from Figure 6-4, this functional unit could be divided into the identification circuit for the contact state and the given circuit for the threshold voltage.

5.2.4 THE UNIT TO GENERATE THE TEST CURRENT

Figure 6-5 shows the block diagram of this circuit.

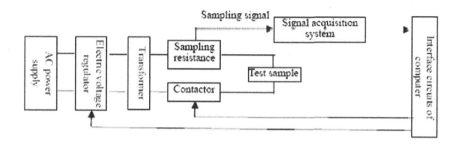

Figure 6-5 The operation principle block diagram of the high-current test cabinet

This functional unit can be divided into two parts, the test current regulating circuit and the test current sampling circuit. Its main function is to provide the required test current for samples according to the standard during instantaneous protection reliability test. The large current test cabinet mainly consists of the large current transformer, the electric regulator and eight contactors, which constitute eight control circuits with eight samples in the sample cabinet through wire connection. To ensure the accuracy of the regulated current, signal sampling and automatic regulation system is equipped in the high-current test cabinet. This system is used for real-time acquisition of current signals, to draw the current curve, and to calculate the effective value of the current in each test; it also used to control the electric regulator to regulate the current automatically until the required current value has been achieved according to the comparison of the calculated value and the required value. The sampling of the current signal and the regulation of the electric regulator are always conducted under computer control.

7

RELIABILITY OF MINIATURE RESIDUAL CURRENT OPERATED CIRCUIT-BREAKER

1 RELIABILITY INDEX

With the continuous improvement of electrification level in industry, agriculture and other industries, the electric power has become one of the indispensable energies in human life and production. With the increase and popularization of electrical equipment, the electrical leakage accidents happen occasionally due to the line fault or improper operation of electrical equipment. Electrical leakage accidents may lead to fire, electrical equipment damage and power personnel injury. To ensure personal and electrical equipment safety, the circuit should be installed with residual current operated circuit breakers (hereinafter referred to as leakage protectors).

The study on leakage protector reliability has attracted increasing attention of the international academic and industry circles. The following can be summarized:

- The reliability verification method has been explicitly specified and issued as early as 1983 by the International Electrotechnical Commission (IEC) in the IEC755 *General Requirements for Residual Current-operated Protective Devices.*
- In 1995, the Chinese national standard GB6829-1995 *General Requirements for Residual Current-operated Protective Devices* was issued and is equivalent to IEC755.

- In 1997, GB16916 *Residual Current Operated Circuit-breakers without Integral Overcurrent Protection for Household and Similar Uses* (RCCB); and GB16917 *Residual Current Operated Circuit-breakers with Integral Overcurrent Protection for Household and Similar Uses* (RCBO) were approved by the National Bureau of Technical Supervision of the People's Republic of China. GB16916 and GB16917 are equivalent respectively to IEC61008-1990 and IEC61009-1990.

Reliability assessment is put forward and the reliability verification is confirmed in all of these standards. Reliability test is specified in GB6829-1995 *General Requirements for Residual Current-operated Protective Devices*, section 7.2: *Performance Requirements*, sub-section 7.2.10.

The reliability testing of leakage protectors was put forward in the above standards. In it, only the severity of the test conditions were specified, i.e., the relatively severe conditions of 28-cycles of repetitive humidity-thermal cycling test. The reliability indices and reliability test method of leakage protectors were not specified, so the reliability tests in the above standards are not true reliability test. The true reliability test should involve a series of contents, such as reliability index, failure criteria, test sampling plan and test results judgment.

In order to promote the reliability development of residual current operated protectors in China, the Chinese national standard GB/Z 22202-2008 *Reliability Test Method of Residual Current Operated Circuit-breakers for Household and Similar Uses* has been issued.

Leakage protectors are one kind of protective apparatus widely applied in low-voltage distribution lines to prevent personal electric shock and leakage fire. The technical performances of leakage protectors mainly include residual current protective performance and operating performance. In case the residual current protective performance of leakage protectors is in excellent conditions, and the electrical equipment is in normal operation without leakage fault, the leakage protectors will not trip.   Once leakage fault occurs, the leakage protectors should quickly separate the contacts and cut off the circuit, to prevent personal injury and fire caused by current leakage. The operating performance of leakage protectors refers to that the circuit can be reliably connected by manually switching ON the leakage protector's handle and can be reliably disconnected by manually switching it OFF. The operating characteristics of leakage protectors are as mentioned above.

Leakage protectors mainly have three fault modes as follows:

(1) Refusing action (Leakage protectors do not OPEN in case of leakage fault)

   The initial performance is lost due to inferior quality of components due to aging and damages caused in service. For example, excessive residual magnetism and serious heat of zero-sequence current transformer; aged permanent magnet; damaged electronic element; poor contact or melting welding of contacts. The refusing action fault is a kind of fault with extreme hazard, leading to the unreliable protection of the person and electrical equipment.

(2) Wrong action (Circuit disconnection due to leakage protector's wrong action without leakage fault)

   In general, residual current exists in circuit, which does not necessarily reflect the fault of the circuits or equipments. For example, leakage current below a certain value is allowed in the qualified capacitors. The insulation resistance of any power supply network and electrical equipment can not be infinite, there is always a certain leakage current in them, so when the value of leakage current is equal to the value of the residual operating current of leakage protectors, the leakage protectors may have wrong action fault. In addition, the leakage protectors may have wrong action fault due to instable structural performance of the leakage protectors, changes in its operating characteristics, influence of load current or ambient temperature, or the electromagnetic interference signals through auxiliary power supply. Wrong action fault leads to the unnecessary power cut of the circuits or electrical equipments, which results in the reliability decrease of power supply and a certain economic loss.

(3) Operation fault

   Leakage protectors cannot be switched on by manual operation, and then the circuit cannot be connected; leakage protectors cannot be switched off by manual operation, and then the circuit cannot be disconnected.

For comprehensive and quantitative reliability assessment of leakage protectors as far as possible, simultaneously taking into consideration the operating characteristics of leakage protectors as a protective apparatus and the three fault modes of refusing action, wrong action and operation fault, two reliability indices are introduced. The grade of protective success ratio R (abbreviated as success ratio) is used as one reliability index pertinent to the refusing action fault and wrong action fault; protective success ratio R refers to the probability that the product can perform the required functions under stated conditions for a specified period of time, or the probability of sucessful testing under stated ocnditions. For operation fault, the grade of operation failure rate λ (abbreviated as failure rate) can be used as another reliability index; failure rate refers to the probability of operation fault occurred in unit time after the moment t when the product has operated until the moment t..

Three failure rate grades (A, B, and C) can be divided according to the value of maximum failure rate λ_{max} ; five success ratio grades (A, B, C, D and E) can be divided according to the value of unacceptable success ratio R_1 . The names of failure rate grades and values of maximum failure rate λ_{max} are shown in Table 7-1, while the names of success ratio grades and values of unacceptable success ratio R_1 are shown in Table 7-2.

Table 7-1   Name of failure rate grades and max. failure rate λ_{max} of leakage protectors

Name of failure rate grades	Max. failure rate λ_{max} （1/10 times）
Grade A	1×10^{-4}
Grade B	3×10^{-4}
Grade C	1×10^{-3}

Table 7-2 Name of success ratio grades and unacceptable success ratio R_1 of leakage protectors

Name of Success Ratio Grades	R_1
Grade A	0.995
Grade B	0.99
Grade C	0.98
Grade D	0.97
Grade E	0.96

2   TEST REQUIREMENTS

The reliability verification test of leakage protectors is recommended to employ the laboratory test.

2.1 ENVIRONMENTAL CONDITIONS

(1) Verification test of success ratio: 28-cycle alternative humidity-thermal cycling test, with cycle as shown in Figure 7-1.

(2) Verification test of failure rate:

Temperature: 15°C to 35°C; relative humidity: 50% to 90%; atmosphere pressure: 86 to 106kPa; avoiding dust and other pollutants. The test samples should be kept for enough time in standard atmosphere (no less than 8h), in order to reaching the thermal equilibrium.

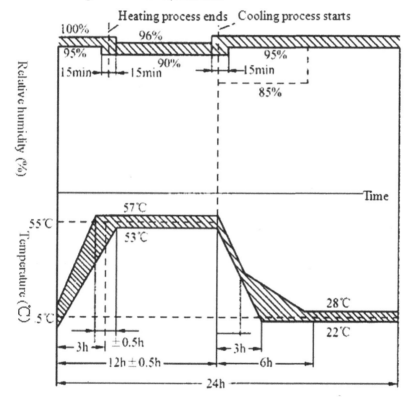

Figure 7-1 Test cycle of success ratio verification test

2.2 INSTALLATION CONDITIONS

(1) The test samples should be installed according to the stipulations of the product standard.

(2) The test samples should be installed in the place without remarkable shock and vibration.

(3) The inclination between the installing surface and the vertical surface of test samples should comply with the product standard.

(4) Leakage protectors that installed by installation rails should be in accordance with the installation rail standards.

2.3 CONDITIONS OF THE TEST POWER SUPPLY

(1) AC power supply should adopt sine wave power with frequency of 50Hz and the allowable deviation of ±5%.

(2) DC power supply may adopt generator, accumulator or stabilized voltage power supply.

(3) When the contact is connected to the load during the test, the voltage fluctuation of test power supply relative to the no-load voltage should not exceed 5%.

2.4 TEST LOAD CONDITIONS

(1) Load power supply can be DC or AC power supply, DC power supply is recommended in general.

(2) The load can be resistive load, sensitive load, capacitive load or nonlinear load, resistive load is recommended in general.

(3) During the test, power voltage U_e in contact loop should be 24V/6V or the minimum DC voltage rating specified in the product standard.

(4) Except that it's specified in standard that the testing current should be rated current, the load current I_c in contact loop during testing should be 1A/100mA.

2.5 EXCITATION CONDITIONS

(1) Verification test of failure rate:

For leakage protectors with $I_e \leq 32A$, the recommended operating frequency is 240 cycles/h, for each cycle, the recommended time interval on off-position is more than 13s; for leakage protectors with $I_e > 32A$, the recommended operating frequency is 120 cycles/h, for each cycle, the recommended time interval on off-position is more than 28s.

(2) Verification test of success ratio:

During the verification test of wrong action fault, energizing current is the rated residual non-operating current; during the verification test of refusing action fault, energizing current is the rated residual operating current.

For the verification test of success ratio, 28-cycle alternative humidity-thermal cycling test is adopted as environmental conditions, complying with the test Db specified in the Chinese national standard GB2423.4-1993 *Basic Environmental Testing Procedures for Electric and Electronic Products — Test*, Db: Cyclic Damp Heat Test Method.

- Test severity degree: 55 ℃ and test cycle of 28 days.
- Description of test cycle is as follows:
- Period 1: Temperature rise process

Gradually raise the testing temperature to the specified upper limit temperature within 3 hours ±30 minutes. The rise rate of temperature should be controlled within the shadowed range as shown in Figure 7-1, during which, the relative humidity should not be less than 95%, and the condensation should appear on the leakage protectors.

- Period 2: Constant temperature 1

Virtually keep the temperature at the upper limit temperature, the specified allowable deviation of ±2°C until 12h±30min (the first half of the test cycle). During this period, except the relative humidity should be between 90% ~ 100% during the first and last 15 min, the relative humidity should be 93%±3% for other period of time. The condensation should not appear on the leakage protectors during the last 15mins.

- Period 3: Temperature drop process

During this process, the temperature should decrease to 25°C±3°C within 3~6h. For different temperature drop process, the decreasing rate of temperature within the first 1h±30min is the same, shown in Table 7-1, if this rate continues to be maintained, the temperature can reach 25°C±3°C within

3h±15min. During the process of temperature decreasing, except that the relative humidity should not be less than 90% during the first 15min, the relative humidity at other time should be no less than 95%.

- Period 4: Constant temperature 2

Keep temperature at 25℃±3℃, and relative humidity of no less than 95% till the end of 24h cycle.

3   TEST METHOD

3.1   PREPARATION OF TEST SAMPLES

All test samples should be randomly sampled from the qualified products that are manufactured by batch production under stable process conditions and screened.

3.2   INSPECTION OF TEST SAMPLES

Test sample inspection can be divided into three items: inspection before test, inspection during test and inspection after test. The inspection during test of leakage protectors can be divided into inspection of failure rate verification test and inspection of success ratio verification test. The test samples without failures during the test must undergo the inspection after test. The inspection items after test include appearance inspection, insulation resistance, dielectric strength and contact resistance, etc.

3.2.1   INSPECTION BEFORE TEST

The inspection before test is to check whether the parts or components of test samples are free of damage and breakage due to transportation. The damaged test samples need to be eliminated and replaced by the well-preserved ones in order to keep the number of test samples required. However, the eliminated test samples should not be counted into the cumulative number of failures r.

3.2.2   INSPECTION DURING TEST

(1) Inspection of Success Ratio Verification Test

1) Inspection of wrong action fault for leakage protectors: during the success ratio verification test, energize the test samples with the rated residual non-operating current to inspect whether the test sample can be in the normal operation;

2) Inspection of refusing action fault for leakage protectors: energize the test samples with the rated residual operating current to inspect whether the test sample can trip within specified operating cycles.

3）Inspection of test button for leakage protectors: press the test button of leakage protectors to inspect whether they can operate.

(2) Inspection of Failure Rate Verification Test: except specified in product standard, otherwise, for each cycle, it needs to detect the contact voltage drop across the closed contact in 40% "switch-on duration" and the open circuit voltage across the contacts in 40% "switch-off duration".

3.2.3 INSPECT AFTER TEST

Except specified in product standard, otherwise, all test samples without failures must undergo the following items after test:

(1) Appearance inspection;

(2) Insulation resistance;

(3) Dielectric strength;

(4) Contact resistance.

3.3 FAILURE CRITERIA

3.3.1 FAILURE CRITERIA FOR SUCCESS RATIO VERIFICATION TEST

During the success ratio verification test, when any of the following cases occurs, the sample is deemed to have one failure:

(1) During the alternative humidity-thermal cycling test, the test sample has wrong action fault after energized with rated residual non-operating current $I_{\Delta no}$.

(2) The test sample fails to trip within the specified operating cycles after energized with rated residual operating current $I_{\Delta n}$.

(3) The test sample fails to trip after the leakage protectors test button is pressed.

(4) The test sample failed to normally close or open because of destructive damage of its parts.

3.3.2 FAILURE CRITERIA FOR FAILURE RATE VERIFICATION TEST

During the failure rate verification test, when any of the following cases occurs, the sample is deemed to have one failure:

(1) Contact voltage drop U_j across the closed contacts exceeds the following limit value U_{jx}. The limit value of contact voltage drop U_{jx} is 5% or 10% of the rated voltage in contact loop.

(2) The open circuit voltage across contacts on open position U_c is lower than the limit value U_{cx}. The value of U_{cx} should be 90% of the rated power voltage of contact loop, except specified in product standard.

(3) Contact welding or sticking.

(4) Leakage protectors cannot close on closing operation command.

(5) Leakage protectors cannot open on opening operation command.

(6) There is destructive damage in the parts or components of test samples, or loose connection between wires and parts.

4 SAMPLING PLAN AND TEST PROCEDURES OF RELIABILITY VERIFICATION TEST

It is suggested to adopt the failure curtailed test for failure rate verification test and MCB success ratio verification test of leakage protectors.

4.1 SAMPLING PLAN OF FAILURE RATE VERIFICATION TEST

The sampling plan of failure rate verification test for different failure rate grades are shown in Table 7-3.

Table 7-3 Sampling plan of failure rate verification test

Allowable number of failures A_C / Truncation time T_r (10^4 times) / Failure rate grade	0	1	2	3	4	5	6	7	8
Grade A	23.0	38.9	53.2	66.8	79.9	92.7	105.3	117.7	130
Grade B	7.68	13.0	17.7	22.3	26.6	30.9	35.1	39.2	43.3
Grade C	2.3	3.89	5.32	6.68	7.99	9.27	10.53	11.77	13.0

4.2 SAMPLING PLAN OF SUCCESS RATIO VERIFICATION TEST

The sampling plan of success ratio verification test at β =0.1 is shown in Table 7-4.

4.3 TEST PROCEDURE OF SUCCESS RATIO VERIFICATION TEST FOR LEAKAGE PROTECTORS

The verification test of success ratio should be conducted according to the following procedure:

(1) To select the success ratio grade.

(2) To select the allowable number of failures A_C .

(3) According to the selected success ratio grade and A_C , to find the number of tests n_f required for acceptance judgment from Table 7-4.

(4) Determination of the number of test samples n . n should not be too small for the sake of test sample representativeness and should be determined according to the size of batches N, the larger N is, the larger n is. And it is recommended that n should be selected from Table 7-5. When determining the value of n, it is also need to take into consideration the two factors: low product price and total test time that should not be too long.

(5) Randomly sample n test samples from the qualified products by batch production, and the size of batches N for sampling should not be less than 10 times that of test samples n , $N \geq 10n$

(6) Determination of the truncated number of test n_z for each sample:

$$n_Z = \frac{n_f}{n} \tag{7-1}$$

(7) Energize the test samples with the rated residual non-operating current, to inspect whether the leakage protectors has wrong action fault.

(8) Energize the test samples with the rated residual operating current, to inspect whether the leakage protectors can trip within the specified operating cycles.

(9) Press the test button of leakage protectors, to inspect whether the leakage protectors can operate.

(10) To make statistics of the total number of failures r_d .

(11) Judgment of test results: In case the accumulated number of tests n_Σ reaches or exceeds the number of tests n_f required by acceptance judgment, while the total number of failures r_d does not exceed the allowable number of failures A_C , the test can be judged to be acceptable; in case the accumulated number of tests n_Σ is less than n_f, while the total number of failures r_d exceeds the allowable number of failures A_C , the test can be judged to be rejectable.

Table 7-4 Sampling plan of success ratio verification test（β =0.1）

A_C / m / Success ratio grade	1	2	3	4	5	6	7	8
Grade A	777	1063	1335	1597	1853	2105	2352	2597
Grade B	388	531	667	798	926	1051	1175	1297
Grade C	194	265	333	398	462	525	587	648
Grade D	129	176	221	265	308	349	390	431
Grade E	96	132	166	198	230	262	292	323

Table 7-5 Recommended value of the number of samples n

Size of batches N	1~2	3~8	9~15	16~25	26~50	51~90	91~150	≥150
Range of n	ALL	2	2~3	3~5	5~8	5~13	8~20	13~32

4.4 FAILURE RATE VERIFICATION TEST PROCEDURE OF LEAKAGE PROTECTORS

The failure rate verification test should be conducted according to the following procedure:

(1) To select the failure rate grade: Grade III or Subgrade IV is recommended for the first test.

(2) To select the allowable number of failures A_C, and A_C is recommended within $1\sim5$.

(3) According to the selected failure rate grade and A_C, find the truncation time T_C from Table 7-3.

(4) To select the test deadline t_z for each sample, $t_z = 6000$ is recommended.

(5) According to T_C, A_C and t_z, determine the number of test samples n by formula (7-2):

$$n = \frac{T_C}{t_z} + A_C \tag{7-2}$$

(6) Randomly sample n test samples from the qualified products by batch production. The number of products for sampling should be not less than 10 times that of test samples n, $N \geq 10n$.

(7) To conduct the test and inspections specified in the above reliability verification test method.

(8) To make statistics of the relevant number of failures r and relevant test time of every failed test sample (failure occurrence time); for the relevant failed test sample inspected after test, its relevant test time is counted as its truncation time at the end of test.

(9) Judgment of test results: In case the accumulated relevant test time T reaches or exceeds the total test time T_C, while the relevant number of failures r does not exceed the allowable number of failures A_C, the test can be judged to be acceptable; in case the accumulated relevant time T does not reach the total test time T_C, while the relevant number of failures r

exceeds the allowable number of failures A_C, the test can be judged to be rejectable.

5 RELIABILITY TEST DEVICE

To study the reliability of low-voltage electrical apparatus, we should not only study its reliability indices and test sampling plan, but also develop the reliability test device to put them into practice. It needs the reliability test device with perfect functions so as to carry the reliability verification tests. To improve the efficiency and correctness of reliability verification test, reliability test device should be developed according to the failure criteria and reliability verification test method mentioned above.

5.1 TECHNICAL FUNCTIONS OF RELIABILITY TEST DEVICE

(1) It has 8 operating mechanisms with fixed installation, capable of performing the failure rate verification test for leakage protectors with single pole or multipole.

(2) According to test requirements, the test personnel can set or modify the related test parameters, such as the operation frequency and the total test time, etc.

(3) During the failure rate verification test, the test device should be capable of simultaneous detecting the open circuit voltage and contact voltage drop of 32-pair contacts for 8 leakage protectors.

(4) During the failure rate verification test, the operation frequency of test samples should be adjustable within $10\sim500$ operations per hour.

(5) It has 2 movable operating mechanisms, capable of performing the success ratio verification test for leakage protectors with single pole or multi-pole.

(6) The test device should generate adjustable sine wave test current within $0\sim500$mA.

(7) Current outputs can be divided into several outputs levels: $0\sim30$mA, $0\sim50$mA, $0\sim300$mA and $0\sim500$mA, and the test device can automatically select the appropriate current outputs level according to the test current settings of leakage protectors.

(8) The test device should possess the function of short-circuit protection, to avoid inter-phase short-circuit after accidental fault of industrial control computer.

(9) During the success ratio verification test, the test device can automatically adjust the test current for any single-pole of multi-pole leakage protectors by software.

(10) The test device can automatically record the test data, such as the number of tests, and in case of failure, can record the serial number of the failed test sample, failure occurrence time and failure mode.

(11) The device possesses the perfect data protection, to avoid loss of data after unexpected power outages, and after power recovery, the acquired data can still be well-preserved.

(12) During the test, in case of test sample failure, the test device can judge whether the test should be stopped or not according to the test parameters; it is easy to perform the test according to the prompts on industrial control computer monitor.

5.2 HARDWARE DESIGN OF TEST DEVICE

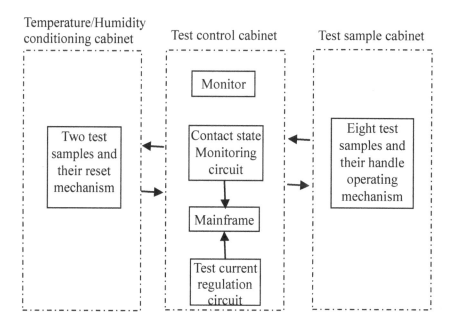

Figure 7-2 Principle block diagram of leakage protectors reliability test device

The reliability test device of leakage protectors is mainly composed of test control cabinet, test sample cabinet and temperature/humidity conditioning cabinet. Its principle block diagram is shown in Figure 7-2. The operation principle of the main units of the test device will be described as follows.

5.2.1 TEST CONTROL CABINET

The test control cabinet is composed of industrial PC, contact state monitoring circuit, test current regulation circuit, input/output card, data acquisition card, digital ammeter and power switch, etc. The main function of contact state monitoring circuit is to acquire the signals of the actual on/off state for all test sample contacts during the switching on/off period after the test sample handle switches on/off in place, and then send them to the computer. Thus, the industrial control computer will judge whether the test sample is normal operation according to the on/off state signals of the contacts.

The test current regulation circuit is composed of step-down transformer, electrical voltage regulator and detection circuit. During the success ratio verification test, the test current regulation circuit is responsible for providing the rated residual operating current and the rated residual non-operating current of leakage protectors. According to the setting value of the test current, the electrical voltage regulator is adjusted to produce sine wave AC current used as the test current in the inspection of leakage protectors.

5.2.2 TEST SAMPLE CABINET

Test sample cabinet is equipped with 8 fixed motor-driven mechanisms to control the closing/opening operation of the test samples; it is capable of conducting the failure rate verification test simultaneously on 8 leakage protectors with single pole or multi-pole; for the leakage protectors with different outline dimensions, the test can be conducted simply by replacing the fixed clips of test samples. The test cabinet is also equipped with 2 movable mechanisms that can be moved into the temperature/humidity conditioning cabinet for the success ratio verification test.

During the failure rate verification test, the switching on/off operation of leakage protectors is completed by the fixed motor-driven mechanism. The forward / reverse rotation of the motors can accomplish the closing/opening operation of test samples' handles. The rotation of the motor is controlled by the industrial computer. The motor should be stopped when the handles of test samples switch on/off in place. Therefore, the operation frequency of test samples can be set or modified according to the test requirement.

During the success ratio verification test, 2 movable mechanisms together with the test samples need to be put into the temperature/humidity conditioning cabinet for severe environment examination of high temperature/humidity, therefore, the movable mechanisms need to go through the corrosion preventive treatment.

The test device can energize two leakage protectors with the rated residual non-operating current, to detect whether leakage protectors have wrong action fault; or energize them with rated residual operating current, to detect whether leakage protectors have refusing action fault. After the leakage protectors trip, the test device can reset it automatically. The test device can perform the inspection of the leakage protectors' test button.

The test device can perform the reliability verification test of the leakage protectors with 1-pole, 2-pole, 3-pole and 4-pole, and can be available for the leakage protectors with different outline dimensions by replacing the clips.

The reliability test device of leakage protectors is equipped with 8 motor-driven mechanisms, each of which has three kinds of manipulators for operating handle, reset button and test button. In case leakage current is more than the rated residual operating current or the test button is pressed, the leakage protectors should trip, the reset button pop up, and the handle jump down. The motor-driven mechanism is capable of automatic reset of leakage protectors through the control of industrial computer. The block diagram of reset button operating circuit is shown in Figure 7-3. The test device can periodically perform the inspection of test button of the leakage protectors through the control of industrial computer. The mechanism of test button is composed of solenoid. The block diagram of test button operating circuit is shown in Figure 7-4.

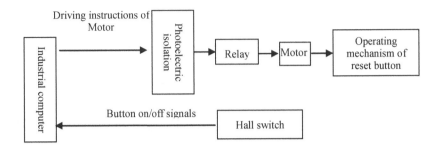

Figure 7-3 Block diagram of reset button operating circuit

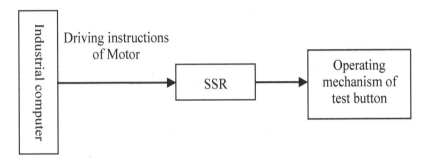

Figure 7-4 Block diagram of test button operating circuit

5.3 SOFTWARE DESIGN OF TEST DEVICE

The application software for leakage protectors' reliability verification test is developed by C language combined with the assembly language. The program can be divided into different modules.

The application software module has excellent human-machine dialogue interface. The software is designed in modular mode. Thus, it is easy for independent use and software debugging, and the test running reliability is also improved. The software contains prompts for operation, for easy operation of the device by test personnel. The test software mainly includes three modules of test inspection, data inquiry and data storage. By data storage module, the test data can be saved in floppy disc. By data inquiry module, the test data can be inquired by test codes.

Test inspection module is for test controlling, test data processing and storage. The test inspection module is the key part of the software, responsible for the storage and display of test data and automatic operations of test device during the test. It includes five parts of test sample testing information, test sample installation, failure rate verification test control program, success ratio verification test control program, and test result display.

The schematic diagram of leakage protectors' reliability test controlling software is shown in Figure 7-5. In the test sample test information program, the information includes the test serial number, test sample label code, the name of test samples' manufacturer, number of test samples, test personnel, test sample model, rated voltage, rated current and so on. The test serial number should be consistent with that in the test agreement, as the identification of different tests. So we use the test serial number as the name of test data file, for test data management and inquiry.

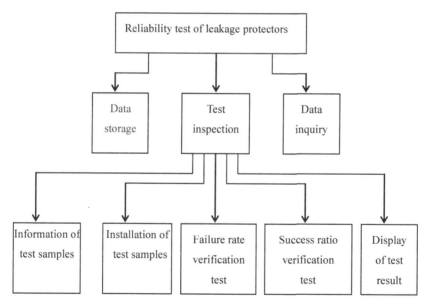

Figure 7-5 Schematic diagram of leakage protectors reliability test controlling software

Test installation module is designed for easily regulating the installation location of test samples and the location of manipulators on the operating mechanism. For the test samples that are manufactured by different manufacturers or those of different models/specifications, the locations of handle, reset button and test button may come with different. If the test sample is installed at improper location, the mechanism will not be capable of correct operation for test sample, and the test result will not be able to reflect the actual reliability level of the test sample. Therefore, before test, the installation location of test samples should be adjusted properly. Test installation module can perform the independent operations of handles of test samples, reset button and test button, for easy adjustment of test sample location.

The flowchart of failure rate verification test controlling program and success ratio verification test controlling program are shown in Figure 7-6 and Figure 7-7. The flowchart of test current adjustment program is shown in Figure 7-8.

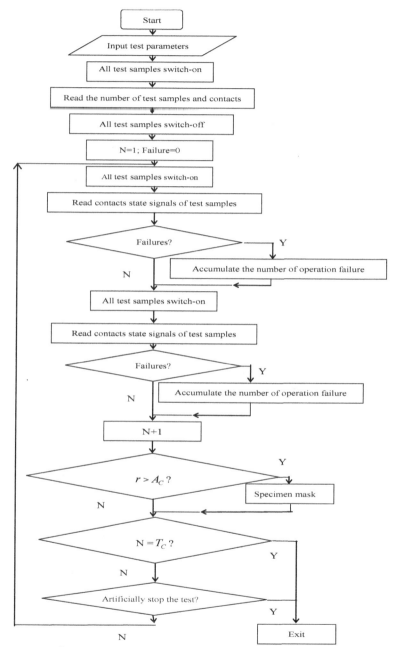

Figure 7-6 Failure rate verification test controlling program

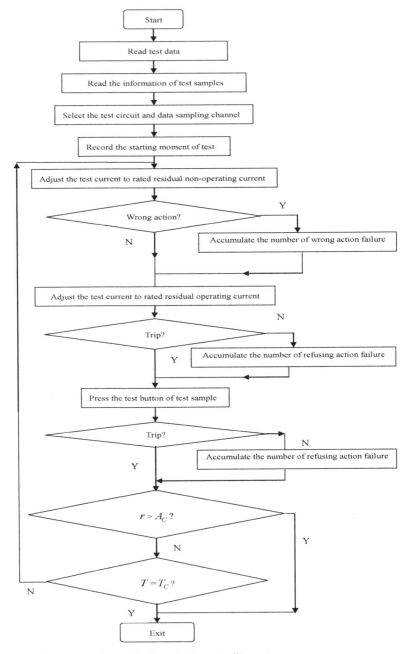

Figure 7-7 Success ratio verification test controlling program

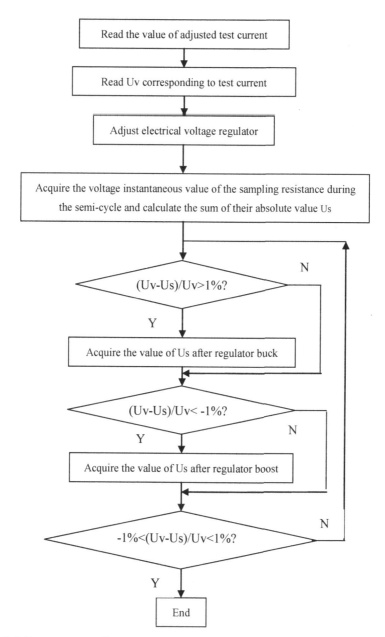

Figure 7-8 Test current adjustment program

8

RELIABILITY FOR LOW-VOLTAGE CIRCUIT BREAKER

1 RELIABILITY INDEX

Reliability working group for low-voltage circuit breaker of IEEE power system reliability committee has conducted the reliability research on the low-voltage circuit breaker and investigated the reliability of it. The failure rates of the release and the calibration of the release are the highest during the survey, which are twice or more higher than the other failure modes. Furthermore, the 13SC group (switchgears) of GIGRE has established a new working group in 1996, that is 13.08 group(the life management of circuit breaker), which was responsible for researching the situation of circuit breaker life management, and aimed to give detailed suggestions on how to extend residual life expectancy of switchgears which are operating or being improved. The research contents included the following items: determination and evaluation for the existing life data and residual life expectancy of high and medium-voltage circuit breaker designed previously or nowadays, the possibility of extending the life and corresponding tests, as well as its economic impacts.

Thus it can be seen that the detailed investigations about reliability of high and low-voltage circuit breakers have been conducted internationally, and people have summarized the major failure modes for them. However, the reliability study involved in the above mentioned literatures were focused on the reliability of circuit breakers which are in use, and only limited to summarize the investigation results, but without any declarative comments. Besides, there were no literatures published about the theoretical contents on the test methods of the circuit breaker reliability verification, as well as the establishment and evaluation methods of the circuit breaker reliability index. The reliability research of circuit breaker in China started relatively late since early 1990s. The reliability management center of Chinese electric power division has issued the accidents investigation and failure statistics and analysis results of high-voltage breakers between 1990 and 1999. The reliability investigation for low-voltage circuit breakers hasn't commenced basically in China.

Through analysis, the major failure modes for low-voltage circuit breakers can be divided into three following types:

(1) Operational fault, it means the low-voltage circuit breaker can not switch on when receiving a closing signal or closed manually, which causes the circuit cannot be connected; The low-voltage circuit breaker cannot switch off when receiving an opening signal or opened manually, which causes the circuit cannot be disconnected.

(2) False operation failure, it means the actions of the instantaneous release and the overload release when there are no overload and short circuit failures with the distribution circuit or the electric equipments, or the actions of the instantaneous release or delayed release caused by the change of operating characteristics of the circuit breaker or the effect of various interference signals, which causes the circuit breaker automatically switches off and thus causes unwanted power outages for the distribution circuits.

(3) Miss operation failure, it means the circuit breaker cannot cut off the failure current reliably in time when there are overload failures or short circuit failures with the distribution circuit, thus causes the electrical circuits or electric equipments cannot be protected reliably.

According to above operating features and failure modes of low-voltage circuit breakers, the operational failure rate level, the success rate level for instantaneous protection and the success rate level for overload protection shall be utilized as its reliability indices. Names of the operational failure rate level of low-voltage circuit breaker and the maximum failure rate, as well as names of the protection success rate level of protection and the unacceptable successes rate refers to Table 6-1 and Table 6-2 in Chapter 6.

2   TEST REQUIREMENTS

2.1 ENVIRONMENT CONDITIONS

Test environment conditions for reliability tests shall be in line with the real working environmental conditions of products. Reliability indices obtained from such tests can reflect the operational performance of this batch of products more truly. However, since the field environmental condition is very complicated and various, it is usually impossible to simulate it accurately during the reliability tests, which is unnecessary on the view of tests.

Thus, unless other provided in the product standards, it is recommended that tests shall be carried out in accordance with the standard atmosphere condition specified in Chinese National Standard GB/T 2421-1999, which is with the temperature of 15℃~35℃, the relative humidity of 50%~90% and the atmospheric pressure of 86~106kPa. Test samples shall be placed under the standard atmosphere condition for such a long period (no less than 8 hours) that the product shall achieve thermal equilibrium. Furthermore, the test environment shall prevent dirt and other contaminations.

2.2 INSTALLATION CONDITIONS

The installation of test samples during the test shall comply with the features of its actual operations. According to the rules of technical standards, the installation shall comply with the following principles:
(1) The test sample should be installed in the place of normal application.
(2) The test sample should be installed in the place without remarkable impact and vibration.
(3) The inclination between the installing surface and vertical surface of the test sample should be consistent with the product standard.
(4) The test sample shall be tested in the free air.

2.3 POWER SUPPLY CONDITIONS

The power supply used for the reliability test of low-voltage circuit breaker can be divided into two types, the DC and the AC. The conditions of AC power supply for test may be established with reference to conditions of electric network source. The two major parameters for evaluating a sine AC power supply are the frequency deviation and the ripple factor. The following items are recommended for use:
(1) The sine AC power supply with the frequency of 50Hz, the allowable frequency deviation is ±5%.

(2) The DC power supply may adopt generator, accumulator or stabilized voltage power supply, or 3- phase full-wave rectification power supply in the case the product performances are not affected in the test, however, its ripple factor shall be no larger than 5%.

(3) Due to the existence of the internal resistance of the test power supply, the load current will be changed when make or break the load during the test, thus the supply voltage applied on the load will be fluctuated. When monitoring the contact reliability of the low-voltage circuit breaker, the accuracy of the measurement will be affected if the contact voltage drop changes too much due to the fluctuation of the power supply voltage. So the fluctuation of the voltage for the test power supply should be no larger than 5% compared to no-load voltage when the contactor makes the load during the test.

2.4 LOAD CONDITIONS OF THE MONITORING CIRCUIT FOR THE CONTACTOR STATUS

(1) Load power supply can adopt DC or AC power supply. Unless specified in the product standard, the DC power supply is recommended.

(2) The load can be resistive load, inductive load, capacitive load or non-linear load. Unless specified in the product standard, the resistive load is recommended.

(3) Unless provided, the voltage of the power supply in the circuit during the test shall utilize 24V or the minimum DC rated voltage of the contactors specified in the product standard.

(4) Unless provided, the load current I_c of the contactor during the test can be 100mA or 1A.

2.5 EXCITATION CONDITIONS OF THE TEST

When performing the instantaneous protecting reliability test for the circuit breaker, the excitation condition of the test should be the test current applied on the test sample, the wave of which should be sine. In accordance with the standard, the test should start with the cold state when verifying the instantaneous protecting characteristic of low-voltage circuit breaker, which means applying 80% and 120% of the short circuit setting current to verify whether false operation failures or miss operation failures occur.

3   TEST METHOD

3.1 PREPARATION OF TEST SAMPLES

The test samples should be randomly sampled from the qualified products produced by batch production under the stable process conditions. The test sample should be new and clean when starting the test.

3.2 INSPECTION OF TEST SAMPLES

The inspection for test samples of the low-voltage circuit breaker can be divided into inspection before test and inspection during the test.

　　During the instantaneous protecting reliability test, start the test with the cold state. Apply 120% of the short circuit setting current and monitor whether the instantaneous release of the test sample acts in 0.2 seconds (operate 10 times). When achieved the cold start, apply 80% of the short circuit setting current, monitor whether the release of the test sample misses the operation, while the duration for the current of the instantaneous release should be within 0.2 seconds (operate 10 times). The instantaneous protecting reliability test should be performed before and after the operational reliability test.

3.3 FAILURE CRITERIA

During the instantaneous protecting reliability test, such test shall be considered to fail when either of the following situations occurs:

(1) The action duration of the instantaneous release for the test sample is no less than 0.2 seconds while applying 120% of the short circuit setting current on the low-voltage circuit breaker.

(2) The action duration of the instantaneous release for the test sample is less than 0.2 seconds while applying 80% of the short circuit setting current on the low-voltage circuit breaker.

4   SAMPLING PLAN AND TEST PROCEDURES FOR THE RELIABILITY VERIFICATION TEST

The verification test plan and procedures for operation failure rate of low-voltage circuit breaker can refer to section 4.1.1 and 4.2.1 in Chapter 6. In this chapter, I will only set forth the verification test plan and procedures for success rate.

4.1 THE VERIFICATION TEST PLAN FOR THE SUCCESS RATE

The test plan for the verification of the success rate under the conditions of four parameters R_0、R_1、α、β set forth in Table 6-6 of Chapter 6 shall guarantee both benefits of users and manufacturers. However, as the number of test samples to make the decision of accepting/rejecting one batch of products is very large, the testing fee is usually very expensive. It's unacceptable for both manufacturers and users during practical applications. Under such circumstance, the test plan, with reference to the determination method of the verification test plan for the success rate of miniature circuit breakers in Chapter 6, another verification test plan which determines the success rate through two parameters R_1 and β can be established, .that is utilizing Table 6-7 as the test plan table for the verification of success rate of low-voltage circuit breaker. Although the form of Table 6-6 differs with table 6-7 a lot, internal relations exist between those two tables. When selecting some R_1 and r_{RE} values from Table 6-7, the verification test plan established from it is the same as the plan figured out from Table 6-6.

For example, In Table 6-6, the test plan determined by $R_0 = 0.99$, $D_R = 3.00$, $\beta = 0.10$ is $n_f = 308$, $r_{RE} = 6$ （$A_C = r_{RE} - 1 = 5$）, where, the R_1 value calclated by the discrimination ratio fomula is $R_1 = 1 - D_R(1 - R_0) = 1 - 3.00 \times (1 - 0.99) = 0.97$. For Table 6-7, the test plan determined by $R_1 = 0.97$, $A_c = 5$ is $n_f = 308$. Thus it can be seen that the verification test plans ascertained by Table 6-6 and 6-7 are just the same.

4.2 VERIFICATION TEST PROCEDURES FOR THE SUCCESS RATE

The verification test of the success rate is based on the assumption that each test is statistically independent. Thus, the test sample must recover the same situation and performance between two adjacent successful tests. Meanwhile, the test results should be divided into two situations, "success" and "failure", when verifying the success rate of products.

The verification test for the success rate shall be performed in accordance with the following procedures:

(1) To select the index of success rate for the product (success rate level).

(2) To select the acceptable failure number A_C

(3) To find out the test number n_f required by the judgement of acceptance from table 6-7.

(4) To designate the test deadline number n_z, usually $n_z = 10$.

(5) To determine the test sample number n according to n_f, n_z and A_C via Fomula (8-1):

$$n = \frac{n_f}{4n_z} + \frac{A_C}{2} \qquad (8\text{-}1)$$

(6) Sample n test samples from the conforming products after screening that are produced by mass production, and the product number for sampling should not be less than 10 times that of the samples n.

(7) Conduct test and detection complying with the test method given in this chapter.

(8) Do a statistics of the total number of failures for all test samples r_d

$(r_d = r_1 + r_2)$; Where, r_1 refers to the number of rejections, r_2 refers to the number of false operations.

(9) Judgment of the test results: When the accumulated test times n_Σ have achieved or exceeded the test times n_f required by making the judgment and the total failure times r_d hasn't achieved the truncation failure times r_C (i.e. $r_d \leq A_C$), then it shall be considered to be accepted by the test (acceptance); When the accumulated test times n_Σ haven't achieved or exceeded the test times n_f required by making the judgment and the total failure times r_d has achieved the truncation failure times r_C (i.e. $r_d > A_C$), then it shall be considered to be rejected by the test (rejection).

5 RELIABILITY TEST DEVICE

5.1 OPERATIONAL RELIABILITY TEST DEVICE

5.1.1 TECHNICAL PERFORMANCES OF THE OPERATIONAL RELIABILITY TEST DEVICE

(1) A bench-style operating floor with three stations, which can operate the reliability tests for three test samples simultaneously.

(2) According to test requirements, users can modify and set various parameters involved in the test.

(3) During the operational reliability test, the test device shall simultaneously monitor the contact drops and the voltages of disconnected contact terminals.

(4) During the operational reliability test, the operation frequency shall be adjustable between 10 and 600 times per hour with an adjustable load factor.

(5) It shall automatically record the test times; when the test sample malfunctions, it shall automatically record the number of the failed test sample, the time of the failure as well as the number of the failed contactor and the type of the failure, furthermore, it should automatically remove the failed test sample. The test data can be handled and printed automatically.

(6) The operating mechanism of the test sample can be driven directly; a manipulator can also be used to operate the handle automatically.

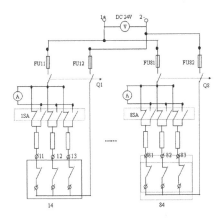

Figure 8-1 Main circuit of the operational reliability test device

5.1.2 HARDWARE DESIGN FOR THE OPERATIONAL RELIABILITY TEST DEVICE

(1) Main circuit. There shall be one wire-wound resistor with 24Ω100W for each test circuit, which shall be with excellent temperature characteristic and won't be influenced by heat. It is as shown in Figure 8-1.

(2) Test excitation control circuit

1) Power-driven operating mechanism. The sketch map for switching on and switching off is as shown in Figure 8-2.

2) Under-voltage release mechanism, the sketch map for the under-voltage release mechanism is as shown in Figure 8-3.

3) Shunt release mechanism, the sketch map for the shunt release mechanism is as shown in Figure 8-4.

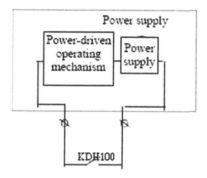

Figure 8-2 The sketch map for power-driven closing

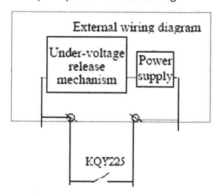

Figure 8-3 The sketch map for the control of the under-voltage release mechanism

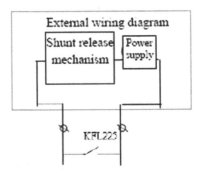

Figure 8-4 The sketch map for the shunt release mechanism

The mechanical operating test procedures for the low-voltage circuit breaker can be easily designed based on above mentioned circuits.

5.2 TEST DEVICE FOR THE OVERLOAD PROTECTION RELIABILITY

5.2.1 TECHNICAL PERFORMANCES OF THE TEST DEVICE FOR THE OVERLOAD PROTECTION RELIABILITY

Overload protection tests for circuit breakers can be divided into three categories:
(1) Cold state test, it means starting performing the overload performance test of the circuit breaker under the cold state.
(2) Hot state test, it means starting the overload performance test of the circuit breaker under the hot state.
(3) Returnable test, it means performing the test on the returnable characteristic of the circuit breaker.

   It should be performed under different overload factors during the test. According to above mentioned sorting methods, the overload protection test can be easily carried out. The test can be divided into the test process and the test cycle on the basis of different stages of the test. One test cycle may compromise one or more test processes, each includes two test currents, which might be generated by different test power supplies, besides, one of the test currents can be zero. During this test, the number of test processes for each test cycle and the test currents for each test process as well as the number of test cycles should be determined in line with test requirements.

For those circuit breakers with no automatic switching on mechanism, the test mode can be set as manual test. Under such manual test, the computer shall wait for the next test of the process automatically after each process test accomplished. After the tester manually closed or opened the test sample, it shall continue the next process test.

The overload protection test device for the circuit breaker shall have the following major performances:

- Users can set the parameters through the software of the test device and such parameters can be modified during the test as well.
- It should be competent to perform tests for three test samples simultaneously.
- The test device shall automatically monitor the test and adjust the test current.
- It shall automatically accomplish the whole test, including the 1.05Ie test, 1.3Ie test and 3.0Ie returnable characteristic test and so on.
- It shall automatically record the test results and shall be able to print the test data.

5.2.2 HARDWARE DESIGN FOR THE OVERLOAD PROTECTION RELIABILITY TEST DEVICE

(1) Computer controlled circuits

The overload protection reliability test device for circuit breakers shall perform the test under the control of the computer. The functional block diagram of the test device is as shown in Figure 8-5. Figure 8-6 is the circuit diagram for the overload protection reliability test device of circuit breakers.

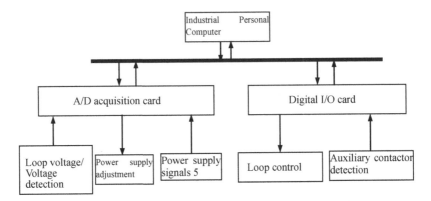

Figure 8-5 The functional block diagram of the test device

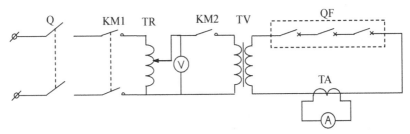

Figure8-6 The circuit diagram for the overload protection reliability test device
Q: knife switch, KM1: contactor, TR: voltage regulator, KM2: contactor, TV:
strong current transformer, TA: current transformer, V: voltmeter, A: ampere
meter, QF: the test sample

There are two power supplies for the test device with the output current
of 0-250A and 0-750A, which shall be used for the overload protection
performance tests. During this test, the test of 1.3Ie shall be performed under
the hot state, which means the test should be performed with the current of
1.3Ie following the accomplishment of the 1.05Ie test under no action of the test
sample. The transition time and setting time between different test currents can
be shortened after duplicate supplies have been employed, which shall improve
the accuracy of test results, and the 3.0Ie returnable characteristic tests will also
be available. However, the returnable characteristic test cannot be performed
with one single supply.

Following the action of the circuit breaker, the test circuit shall
disconnect, thus in general, one test power supply shall only serve tests of one

test sample. The overload characteristic test should last 2-4 hours, so it shall take you a long time to accomplish tests for multiple test samples. To speed up the test, such multiple test samples can be connected in series, which makes it possible to perform tests simultaneously. After the action of one or multiple test samples, the circuit shall automatically switch over to the balanced loop and continue the test for other test samples. The internal resistance of the balanced loop is very close to the test sample, meanwhile the current shall automatically stabilized through the computer, thus the change of the test current will be very tiny and the duration for such change will be very short as well.

(2) Current regulating circuit

Since the variation range of current during the overload protection reliability test is very wide, to improve the accuracy of control and detection, an appropriate current transformer should be selected in accordance with the test current magnitude. For example, for a circuit breaker with the rated current from 10A to 225A, the test current can be divided into four ranges, including 0~25A, 0~75A, 0~250A and 0~750A. The electrical block diagram for the regulation of the test current is as shown in Figure 8-7.

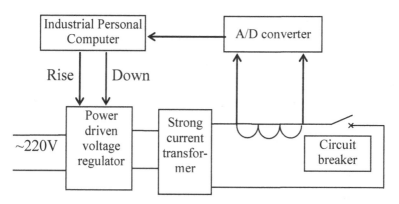

Figure 8-7 Electrical block diagram for the regulation of the test current

The regulation of the test current is performed under the control of the computer. The operator sets the current parameters and the computer will automatically regulate and stabilize the current as required. When the variation of the test current is caused by the voltage fluctuation of the power supply and the variation of the loop resistance, the computer will automatically regulate the current to the setting value; such current regulation is with high precision and fast response speeds, which will grantee the accuracy of test results. Both the test for one single test sample and for multiple test samples can be performed by selecting the test loop.

(3) Detection circuit for the contactor status

The major function for the detection of the contactor circuit is to judge the statuses of the auxiliary contact and the main contact for the circuit breaker, i.e. judging the action situation of the circuit breaker under the test current. This circuit will send the signal to the industrial computer under optoelectronic isolation.

5.2.3 SOFTWARE DESIGN FOR THE OVERLOAD PROTECTION RELIABILITY TEST DEVICE

(1) Some significant issues for test program design under Windows

1) Real time requirements. The hardware and software refers to two important aspects of the application development system. The hardware is the foundation while the software is the key. The construction of the hardware circuits provides necessary means and possibility to accomplish functions of the device, while the accomplishments and reliable operation of all functions is mainly based on the design of the software. The software for the overload protection reliability test device is developed on the basis of Visual C++6.0 under Windows2000. As a high level language, VC provides interface commands for the connection with the hardware, thus the direct control of the hardware can be achieved while a complicated human-computer interface program can be expediently compiled.

The software of the overload protection reliability test of the circuit breaker which demands to be processed in real time differs from common applications of Windows. To grantee the accuracy of the time and the test current, the accuracy of the software timing shall be taken into account. To grantee the real time control, both the hardware delay and the software delay should be employed in the software. The hardware delay shall be used under the situation of shorter delay time and higher delay accuracy, while the time control delay shall be used under the situation of longer delay time and lower delay accuracy.

2) Running of single instance application. The application is able to run simultaneously for multiple times under Windows operation system. In the

overload protection reliability test of circuit breaker, the interrupt mode is employed for data acquisition, thus occupies the hardware resource of the computer and external hardware resource. If the software is under running, running the implication repeatedly will lead to conflicts of hardware resource and cause the program cannot function normally. To prevent such false operation, the simultaneous running of multiple instances should be banned when programming.

3) Modular Design. To facilitate the operation of the test, various functions will be designed when writing the software of the overload protection reliability characteristic test of the circuit breaker. Some of such functions are very similar, so modular design should be employed for the control software. It not only facilitates multiple use and software debugging, and also enhances the reliability of the test. When writing the program, the operation hints for testers should be taken into account, and an excellent human computer interface should be equipped to facilitate the testers to operate the test devices.

(2) Design of the test program

The program of the overload protection reliability test mainly consists of four major modules, the test run, the data display, the storage and access of data, parameter setting. The test run is the key portion of this software, which is in charge of the running of the whole test and storage of test data. It mainly includes operation module for single test, operation module for cyclic tests, regulating/stabilized current control module. The procedures for the subprogram of single test are as shown in Figure 8-8, when the procedures for the cyclic tests subprogram are as shown in Figure 8-9.

During the overload protection reliability test, the heat generation of the conductor will influence the overload characteristic when regulating the current from 0 to the test current, the longer the regulating time, the larger the influence . To regulate the current quickly, the method employed for current regulation is as follows: when the current arise from zero to a certain percentage of the test current, such as 20% or 50% of the test current, disconnect the test loop and measure the voltage. Regulate the voltage as required on the basis of the test current, for example, if the current is regulated to be 20% of the test current, then increase the voltage to 5 times, if the current is firstly regulated to be 50%, then increase the voltage to 2 times, then close the test loop and stabilize the current to achieve fast regulation for the test current, while the influence to the overload protection test results caused by heat generation produced by the current regulation will be greatly reduced. The flow chart for the current regulation is as shown in Figure 8-10.

The data display function is mainly used to check the test results to make sure users can get to know the test situations of each test sample.
The data storage and access module may save the test result as other documents to facilitate long term preservation of the teat result for users. When performing new tests, the original test data shall also be saved in other documents. The original data file can be loaded into the computer and check and print such test data.

Test parameters setting module is mainly used for modifying and setting the test parameters. Users can modify and set the test parameters on the basis of test parameters for different test samples.

5.3 INSTANTANEOUS PROTECTING RELIABILITY TEST DEVICE

5.3.1 EXISTING PROBLEMS OF THE INSTANTANEOUS PROTECTING RELIABILITY TEST DEVICE

According to related rules in Chinese National Standard GB14048.2-2001 *Low-voltage Switchgear and Control Gear Low-voltage Circuit Breakers*, the release limit and characteristics shall comply with the following rules: when the test current is equal to 80% of the short circuit setting value, the release shall not act; when the test current is equal to 120% of the setting value of the short circuit current, the release shall act. For the instantaneous release characteristic test of multi-pole circuit breakers, the two poles of test electrical appliances shall be connected in series. When the release or rejection of release test is performed with the test current, the current duration in the instantaneous protecting reliability test should be 0.2 seconds.

The instantaneous protecting reliability test for circuit breakers is normally performed on the instantaneous calibration board. The main circuit for the instantaneous calibration board is as shown in Figure 8-11.

The inaccuracy of the setting current of the circuit breaker calibrated by the instantaneous calibration board is mainly caused by the following aspects.

(1) The voltage fluctuation of the power grid. From Figure 8-11 it shall be seen that the factory calibrate with the open-charging- voltage means, thus if the voltage of the power grid fluctuates, the test current will be influenced, which brings about errors into the calibration results.

(2) The inconformity of internal resistances of test samples. When calibrating circuit broakorc with the came specification, usually the value of the peak value meter for the first circuit breaker is measured, while the current for the test circuit will not be checked. Such verifying method is impropriate, due to the material and technical process, the internal resistances of each circuit breaker, even the internal resistances for different poles of the same circuit breaker differ, especially for those circuit breakers with lower rated current, the differences between each pole is more significant. Thus, the current value actually passing the circuit breaker differs, so errors of the setting value still exits for circuit breakers calibrated in such manner.

In Formula 8-2, the first item on the right side refers to the periodic component; the second item refers to a damped aperiodic component. It can be seen that the aperiodic component is connected with the voltage phase angle of closing ψ. Due to the difference of phase angles of closing and the random test currents with different values, the actual setting value is not uniform, which may cause the action under condition of 80% of the setting value or the rejection of action under the condition of 120% of the setting value. Thus, the aperiodic component of the circuit current during the closing instant greatly affects the current setting value of the circuit breaker.

(3) Variation of the test loop resistance. Room temperature may be changing during the test, which shall lead to the resistance of the test loop. In addition, the resistance of the test loop is under the cold state at the beginning of the test, while after a period of test, the test loop will be under the hot state. Thus the change of temperature will lead to the change of the resistance of the test loop, which shall affect the setting value of the test current.

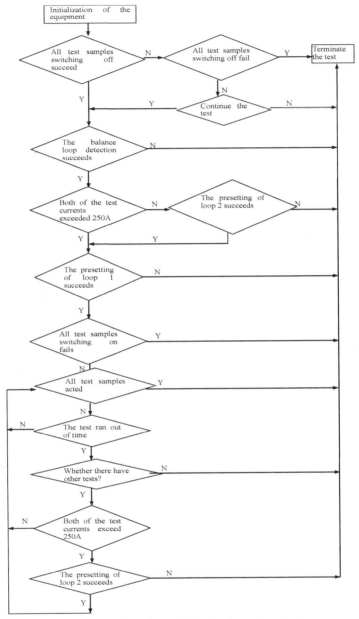

Figure 8-8 The subprogram flowchart of the single test control

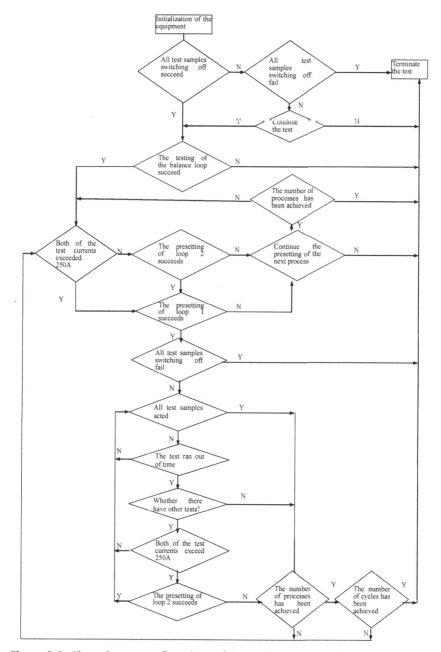

Figure 8-9 The subprogram flowchart of the cyclic test control

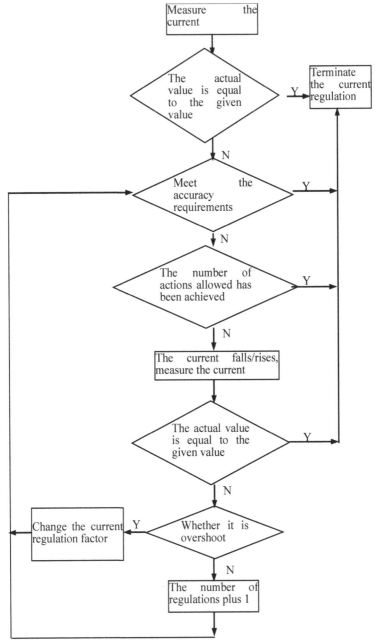

Figure 8-10 The control flowchart of the current regulation/current stabilization

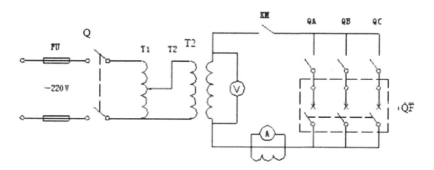

Figure 8-11 The main circuit of the instantaneous calibration board. FU: fuse, Q: knife switch, T1: automatic coupling voltage regulator, T2: strong current transformer, KM: contactor, QA,QB,QC: knife switches, V: voltmeter, A: ampere meter, QF: the test sample

5.3.2 TRANSIENT PROCESS ANALYSIS FOR THE INSTANTANEOUS PROTECTING RELIABILITY TEST

5.3.2.1 TEST CIRCUIT

The test circuit requirements of the instantaneous protecting reliability test is as follows: for circuit breakers used for power distribution, the transient calibrating current provided by the test circuit shall achieve 10 I_N with the error of $\pm 20\%$. For circuit breakers used for the control of motors, the transient calibrating current provided by the test circuit shall achieve 12 I_N with the error of $\pm 20\%$; besides the test circuit shall have the off-timing function. Currently, the common basic circuit diagram for the instantaneous characteristic test of the circuit breaker is as shown in Figure 8-12.

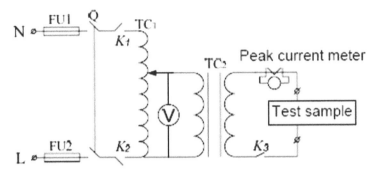

Figure 8-12 Basic circuit diagram for the instantaneous characteristic test of the circuit breaker. FU: fuse, Q: knife switch，TC1: self-dual voltage regulator, TC2: strong current transformer, K1: contactor，K2: delay relay, V: voltmeter, TA: peak current mete, QF: the test sample

This circuit consists of the self-dual voltage regulator TC1, the strong current transformer TC2, the Control switching on contactor K1 and the delay breaking time relay K2. Utilize the delay breaking function of the normally closed contact of the time delay to break the circuit at regular time.

Table 8-1 Instantaneous characteristic requirements of circuit breakers

Circuit breakers used for power distribution	Current applied	$8I_N$	$12I_N$
	Release time	$\geq t_T$	$\leq t_T$
Circuit breakers used for the control of motors	Current applied	$9.6I_N$	$14.4I_N$
	Release time	$\geq t_T$	$\leq t_T$

According to the release requirements of the circuit breaker, the test sample shall be judged to be accepted or rejected in accordance with Table 8-1. In this table, t_T refers to the time limit of release specified by the product standard. It's usually 100ms~200ms on the basis of different requirements for different models of products. The detailed test method is as follows:

(1) Connect the test sample and close K1, regulate the current of the test loop to be I_N through the voltage regulator, record the output voltage U_N of the voltage regulator and cut off K1.

(2) Regulate the output of the voltage regulator to be $8(9.6)kU_N$, herein, $k > 1$ and the value of which relates to the internal resistance of the strong current transformer. k Shall be determined through experiences to ensure the current of the test loop achieves $8(9.6)I_N$.
Close K1 and connect the circuit and check whether the test sample releases.

(3) The phase angle of closing differs. When performing the instantaneous characteristic test, the AC voltage is applied randomly, and the instantaneous current value passing the release coil refers to

$$i = I_m \sin(\omega t + \psi - \varphi) - I_m \sin(\psi - \varphi)e^{-\frac{\omega t}{tg\,\varphi}} \qquad (8\text{-}2)$$

Where, ψ = the voltage phase angle of closing, φ—the power factor angle of the test circuit, I_m = the peak value of the current periodic component.

(4) Regulate the output of the voltage regulator to be $12(14.4)kU_N$, the selection principle of k is idem. Close K1 and connect the circuit to check whether the test sample releases.

(5) Judge whether the test sample is qualified on the basis of the release situation of the test sample.

From above mentioned test methods it can be seen that applying the current $8(9.6)I_N$ and $12(14.4)I_N$ on the test sample is achieved through applying the voltage $8(9.6)kU_N$ and $12(14.4)kU_N$ on the test sample. It's just because the test sample might be caused to release when applying the current $8(9.6)I_N$ and $12(14.4)I_N$ on the test sample of the circuit breaker. There is no time for the test device to guarantee the accuracy of the current thorough certain control measures, thus the control of the voltage is the only way to indirectly guarantee the accuracy of the current. The control principle can be expressed as Formula (8-3) and Formula (8-4).

$$U_N = I_N Z \qquad (8\text{-}3)$$

$$8(9.6)I_N = \frac{8(9.6)kU_N}{Z^*} \qquad (8\text{-}4)$$

Determine U_N corresponding to I_N from Formula（8-3）, Z refers to the resistance of the test loop, including the contact resistance of the circuit breaker, the release detection circuit, and the impedance of the regulator and transformer.

When applying $8(9.6)I_N$ on the test sample, due to the position change of the voltage regulator, the loop impedance in Formula (8-4) will be changed to Z^*, generally $Z^* > Z$, thus the voltage applied on the test sample should be $8(9.6)kU_N$. To guarantee that the test loop produces $8(9.6)I_N$, k should satisfy Formula (8-5).

$$k = \frac{Z^*}{Z} \qquad (8\text{-}5)$$

Due to the change of the impedance with no certain rules caused by the change of the regulator position, the value of k can only be estimated manually through experience.

From the steady state it can be seen that the accurate test current can be obtained during test through accurate estimation of k value and the assurance of the voltage accuracy. However, for the transient state, it's more difficult to obtain the accurate test current.

When the inductance load or the capacitive load exists in this circuit, the current passing through the contact at the closing moment includes the steady state component and the transient state component, herein, the steady state component refers to the periodic component, while the transient component also refers to the aperiodic component. Due to the existence of the aperiodic component, the peak value of the current is larger than theoretic value, which may cause the unexpected release of the qualified product when applying the $8(9.6)I_N$, and may cause the release of the rejected product when applying the $12(14.4)I_N$, thus lead to the detection errors of the instantaneous characteristic of circuit breakers. So the elimination of the effect the transient current (the aperiodic current) to the test is very important.

5.3.2.2 TRANSIENT CURRENT

As the aperiodic component of the current during the test influences the test results, it should be eliminated.

(1) Generation of the periodic component of the current. To working out the reason for the generation of the aperiodic component of the current, It Is the priority to analyze the main circuit of the test. The equivalent circuit of Figure 8-12 is as shown in Figure 8-13.

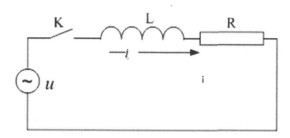

Figure 8-13 The equivalent circuit diagram of the test circuit

At the instant of closing the switch K, the voltage equation of the circuit is

$$u = Ri + L\frac{di}{dt} \tag{8-6}$$

Where, u refers to the power supply voltage. Assume the power supply voltage at the instant of switching on refers to

$$u = U_m \sin(\omega t + \psi) \tag{8-7}$$

Substitute Formula (8-7) into Formula (8-6) and we will get:

$$i = \frac{U_m}{Z}\sin(\omega t + \psi - \varphi) - \frac{U_m}{Z}\sin(\psi - \varphi)e^{-\frac{t}{T}} \tag{8-8}$$

Where, $Z = \sqrt{R^2 + (\omega L)^2}$ refers to the loop impedance; $\varphi = \arccos\dfrac{R}{Z}$ refers to the power factor angle; $T = \dfrac{R}{L}$ refers to the time constant of the circuit.

Mark the $\dfrac{U_m}{Z}\sin(\omega t + \psi - \varphi)$ as i_1, and $\dfrac{U_m}{Z}\sin(\psi - \varphi)e^{-\frac{t}{T}}$ as i_2.

From Formula (8-8) it can be seen that i_2 is a component that decays according to the exponential law along with time and shall tend to be 0 with the increase of time. This component is an aperiodic component. i_1 is a periodic function of time with the effective value of $I_1 = \dfrac{U_m}{\sqrt{2}Z}$, which is the anticipated test current. ψ is the voltage phase at the instant of closing, which directly influences the aperiodic component of the test current.

(2) The influence of the aperiodic component of the test current

The current produced at the instance of closing consists of two components, i_1 and i_2. i_1 is a periodic component, which is the anticipated test current for users, while i_2 is a aperiodic component which is a transient process and an unwanted component during the test. The following paragraphs will analyze its influence on the test.

Firstly discuss an extreme case:

$$\psi - \varphi = -\frac{\pi}{2} \tag{8-9}$$

Substitute $\psi - \varphi = -\dfrac{\pi}{2}$ into i_1, i_2 and i and we will get:

$$i_1 = -\frac{U_m}{Z}\cos\omega t \tag{8-10}$$

$$i_2 = -\frac{U_m}{Z}e^{-\frac{t}{T}} \tag{8-11}$$

$$i = \frac{U_m}{Z}(e^{-\frac{t}{T}} - \cos\omega t) \tag{8-12}$$

Figure 8-14 shows the contrast of waves of i and i_1. The solid line curve refers to the wave of i, while the dash line curve refers to the wave of i_1.

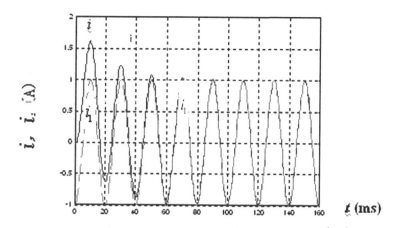

Figure 8-14 The test current waves for the instantaneous characteristic of the circuit breaker

From Figure 8-14 it can be seen that the peak value of i is 60% larger than i_1, which means the test current actually produced is 1.6 times of the required value. According to this multiplying power, the current actually produced is 12.8(15.4)/I_N while our required current is 8(9.6)/I_N Thus most circuit breakers will release under such current.

Figure 8-15 gives a relationship between the voltage angle of closing and the peak value of current in the instantaneous characteristic test of the circuit breaker. In this figure, the horizontal ordinate refers to $\psi - \varphi$ with the unit of radian, while the vertical coordinate refers to the peak ratio of i and i_1. From this figure we can see that when $\psi - \varphi = \pm\dfrac{\pi}{2}$, the distance between the current produced and the anticipated current is the largest. While $\psi - \varphi = 0$, the current produced is just equal to the anticipated current.

In summary, when performing the instantaneous protecting reliability test for the circuit breaker, the aperiodic component produced by the current relates to the phase of the power supply at the instance of closing, which means it relates to the phase angle of closing. When the phase angle of closing is equal to the power factor angle of the test circuit, no aperiodic component of current will be produced.

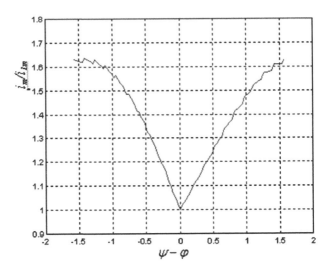

Figure 8-15 The relationship between the phase angle of closing and the peak value of current

To eliminate the aperiodic component of the current, the phase angle of closing is required to be equal to the power factor angle of the test circuit. For this purpose, the power factor angle should be measured firstly. To achieve a better accuracy of measurement, the fast Fourier transforms shall be employed. Based on the signals of voltage and current of the test loop measured, calculate the power factor angle of the circuit, and on this basis, perform the plan of phase selection.

5.3.2.3   PRINCIPLES AND METHODS OF ELIMINATING THE NON-PERIODIC COMPONENTS OF TEST CURRENT

It can be seen from the above analysis that the non-periodic components of test current can be eliminated when 0. Therefore, the most important thing is to rapidly obtain the accurate value of power factor angle cp of test circuit. It is accomplished by the Fast Fourier Transform (FFT) method.

(1) Fast Fourier Transform method

There are two methods of signal analysis: time-domain analysis and frequency domain analysis. FFT is used to convert the time-domain signal into a frequency domain signal. FFT is shown in Eqs.(8-13) and (8-14), where is the time-domain signal, $H(f)$ is the frequency domain signal.

$$H(f) = FT(h(t)) = \int_{-\infty}^{+\infty} h(t) e^{-j2\pi ft} dt \qquad (8\text{-}13)$$

$$h(t) = FT^{-1}(H(f)) = \int_{-\infty}^{+\infty} H(f) e^{j2\pi ft} d \qquad (8\text{-}14)$$

The above shows the mutual conversion between the time-domain signal and the frequency- domain signal. This method can be used to analyze the spectrum of the time-domain signal and similarly, represent the time-domain signal according to the signal spectrum.

$H(f)$ is a complex function, which contains real and imaginary parts: $H(f) = \operatorname{Re} H(e^{j\omega}) + \operatorname{Im} H(e^{j\omega})$. It can be also expressed in amplitude and

phase : $H(f) = |H(f)| e^{j\phi(f)}$, where, $|H(f)| = \sqrt{(\operatorname{Re} H(e^{j\omega}))^2 + (\operatorname{Im} H(e^{j\omega}))^2}$,

$\phi(f) = \tan^{-1}\left[\dfrac{\operatorname{Im} H(e^{j\omega})}{\operatorname{Re} H(e^{j\omega})}\right]$.

In fact, the sampling of the signal is always conducted in a limited time interval, which is called as time window. Fig.8-16 shows the available sampling data and the spectral characteristics of their counterparts in the limited time.

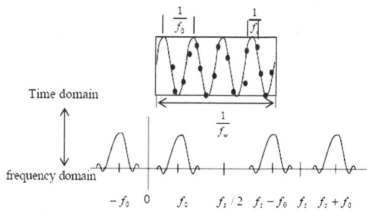

Figure 8-16 The sampling data and the spectral characteristics in time window

There are several definitions of time intervals: the given component cycle of the spectrum $1/f_0$, the cycle of time window $1/f_w$, the sampling cycle $1/f_s$.

Their radio frequency domain characteristics arc: spectrum frequency f_0, peak width f_w, spectrum cycle interval f_s.

By the Z Transform of Eqs.8-13 and 8-14, the Discrete Fourier Transform (DFT) can be obtained:

$$X(k) = \sum_{n=0}^{N-1} x(n) e^{-j\frac{2\pi nk}{N}} \quad (k = 0,\ 1\ \cdots\ N-1) \tag{8-15}$$

$$x(n) = \frac{1}{N} \sum_{k=0}^{N-1} X(k) e^{j\frac{2\pi nk}{N}} \quad (n = 0,\ 1\ \cdots\ N-1) \tag{8-16}$$

Where N is the number of data acquisition in time window in the time domain, i.e.,

$$N = \frac{f_s}{f_W} \tag{8-17}$$

$x(n)$ is N sampling data in the time domain, $X(k)$ is K data in the discrete spectrum. The number of discrete spectrum is determined by the number of sampling points in the time domain, i.e., the resolution of spectrum v is:

$$v = \frac{f_s}{N} = f_w \tag{8-18}$$

(2) Calculation of power frequency

The power frequency should be calculated first in order to obtain the early phase angle of the test voltage and current signals. The specific methods of calculating the frequency is: put N sampling data of voltage into arrays $x(n,1)$, FFT, then put the real part of results into arrays $x(n,1)$, and the imaginary part into arrays $x(n,2)$; identify the absolute best location k of the real and imaginary parts. The power frequency can be calculated by the following formula: $f_0 = \dfrac{(k-1)f_s}{N}$

(3) Calculation of power factor angle

The power factor angle is the phase difference between voltage and current in the same loop. Thus, the power factor angle can be obtained by calculating the early phase angle based on the analysis of sampling data of voltage and current. The voltage and current signal can be expressed as:

$$h(t) = \sin(2\pi f_0 t + \alpha) \tag{8-19}$$

Where a is the early phase angle of voltage or current signal.

If $f_0 = Mf_w$ is not valid, the approximate calculation method of the early phase angle is:

Supposed that $t_W = Mt_0 + t_\Delta$, i.e. t_Δ is the remainder of to in addition to tw.

Assume that $\theta = 2\pi f_0 t_\Delta$, FFT is equivalent to:

$$H(f_0) = \int_{-t_W/2}^{t_W/2} \sin(2\pi f_0 t + \alpha + \frac{\theta}{2})e^{-j2\pi f_0 t} dt \qquad (8\text{-}20)$$

Assume that $\beta = \alpha + \dfrac{\theta}{2}$, then Eq.(8-20) will be:

$$H(f_0) = \int_{-t_W/2}^{t_W/2} \sin(2\pi f_0 t + \beta)[\cos(2\pi f_0 t) - j\sin(2\pi f_0 t)]dt$$

$$= \frac{1}{4\pi f_0}\{\sin\theta\sin\beta + (2\pi M + \theta)\sin\beta - -j[\sin\theta\cos\beta + (2\pi M + \theta)\cos\beta]\}$$

Because $(2\pi M + \theta) \gg \sin\theta$, the above formula can be approximated to:

$$H(f_0) \approx \frac{2\pi M + \theta}{4\pi f_0}(\sin\beta - j\cos\beta) \qquad (8\text{-}21)$$

It can be concluded from Eq.(8-21) that the ratio of real and imaginary parts of spectrum in fo is:

$$\frac{\operatorname{Re}H(f_0)}{\operatorname{Im}H(f_0)} = -tg\beta \qquad (8\text{-}22)$$

Because $\beta = \alpha + \dfrac{\theta}{2}$, then

$$\alpha = -arctg\frac{\operatorname{Re}H(f_0)}{\operatorname{Im}H(f_0)} - \frac{\theta}{2} \qquad (8\text{-}23)$$

When $f_0 = Mf_w$ is valid, then $t_W = Mt_0$, $\theta = 0$

$$H(f_0) = \frac{M}{2f_0}(\sin\alpha - j\cos\alpha) \qquad (8\text{-}24)$$

The early phase angle is:

$$\alpha = -arctg\frac{\operatorname{Re}H(f_0)}{\operatorname{Im}H(f_0)} \qquad (8\text{-}25)$$

Thus, the power factor angle φ can be obtained by calculating the early phase angle of voltage and current.

5.3.3 THE TECHNICAL PERFORMANCES OF THE INSTANTANEOUS PROTECTING RELIABILITY TEST DEVICE

(1) Eliminate the influence to the instantaneous characteristic commissioning caused by the aperiodic component produced by the traditional instantaneous characteristic commissioning unit of circuit breaker

(2) The commissioning unit is competent to capture the strong current required by the test quickly and accurately.

(3) The commissioning unit shall be equipped with multiple debugging and commissioning tap positions. The test sample can be connected to the corresponding checking tap position as required by the commissioning test.

(4) As required by the test, testers can input parameters through the computer interface such as the value of rated current for the test sample, the multiple of the lower limit and the upper limit of the current, as well as the release time and so on.

(5) Complete data protection function, such as no data loss during an unexpected outage and no damage to the collected data after the recovery of power.

(6) The test device can automatically record the test results and can print and output the data and waves detected.

(7) During this test, the computer shall perform operation hint and reduce false operations.

5.3.4 HARDWARE DESIGN FOR THE INSTANTANEOUS PROTECTING RELIABILITY TEST DEVICE

(1) Constitution of the hardware. The instantaneous protecting reliability test device for the circuit breaker comprises of two major parts, the computer control cabinet and the operation cabinet. The functional block diagram for the unit is as shown in Figure 8-17. The computer control cabinet comprises of industrial PC, the detection circuit and control circuit and so on. The operation cabinet mainly comprises of the voltage regulator and the strong current transformer. There are two wiring terminals for this commissioning unit, and is competent to connect one test sample.

The computer control cabinet   Operation cabinet

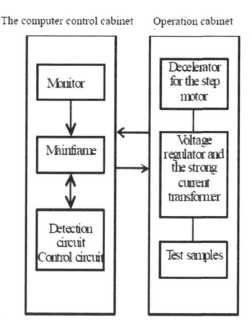

Figure 8-17 The functional block diagram for the instantaneous protecting reliability test device

The major function of the detection circuit for the test loop is to detect the voltage, current, power factor angle and the voltage phase for the test loop. This circuit will send the signal to the industrial computer under optoelectronic isolation.

The control loop mainly comprises of the contactor, the control delay and the solid state relay, among which two control relays are used for the control of the positive and negative rotations of the stepping motor to increase or reduce the current of the test loop.

(2) Hardware circuit. The hardware circuit for the instantaneous protecting reliability test device of the circuit breaker includes two major portions, the main loop, the computer controlled circuit and detection circuit. The main loop provides the current for the instantaneous protecting reliability test. The computer controlled and detection portion is mainly in charge of functions such as the control of the timing sequence, the measurement of the test parameters, the display of the test waves and values, receiving the setting values of parameters by users, the judgment of approved condition of the test sample and the storage and processing of the data and so on.

1) Design of the main loop for the test device. When performing the instantaneous protecting reliability test for the circuit breaker, the requirements for the current applied and the release is as shown in Table 8-1. The major circuit of the commissioning unit is as shown in Figure 8-18. This circuit mainly comprises of the power driven automatic coupling voltage regulator TC_1, the AC solid state relay ACSSR, the strong current transformer TC_2, the current detection resistance R1, the balance resistance R2 and several control contactors J1~J5.

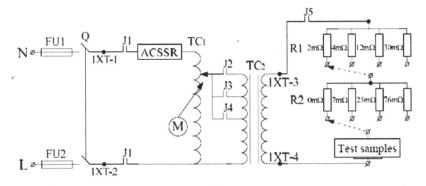

Figure 8-18 The major circuit for the test device

The power driven voltage regulator TC_1 achieves the regulation of the test current thorough the control of the motor rotation. When the motor is under positive rotation, the output voltage of the voltage regulator will rise. When the motor is under negative rotation, the output voltage of the voltage regulator will fall. The primary side of the strong current transformer TC_2 has three taps, which is connected with J4, J3 and J2 respectively. When performing the test for the molded case circuit breaker, firstly close J2 and J5. Make the current of the test loop achieve the rated current through the regulation of the power driven voltage regulator. Suppose the no-load voltage between 1XT-3 and 1XT-4 is U_N, when J2 opens and J3 closes, the no-load voltage between 1XT-3 and 1XT-4 will be $8U_N$; when J2 closes and J3 opens, the no-load voltage between 1XT-3 and 1XT-4 will be $12U_N$. Then by cooperating with the regulation of the power driven voltage regulator to eliminate the voltage errors caused by the inaccurate ratio of transformation, we can quickly and accurately callout the required test currents $8I_N$ and $12I_N$. When I_N is 6A~225A, the current needed to be detected is 6A~3240A. The detection range is very large. To improve the detection accuracy, the current detection resistance R1 is divided into 4 tap positions. The test sample is connected to the corresponding level according to the rated current of the test sample. The function of the balance resistance R2 is to make the power driven voltage regulator TC_1 operate in the high voltage area to relatively improve the regulation accuracy. The corresponding current detection resistance R1 of the balance resistance is divided into 4 tap positions. The calculation for parameters is as follows:

a) The detection is divided into 4 levels in accordance with the rating current of the circuit breaker. The levels for detection are as shown in Table 8-2.

b) Design of the selection of the detection resistance R1: Generally speaking, the range of the detection signal for AD acquisition board is ±10V, the ratio between the AC peak value and the effective value is $\sqrt{2}$. Thus the range of the effective value for the AC voltage detected is ±7V. The current detection range for each loop is 14.4 times of the range between the lower limit current and the upper limit current. Table 8-3 shows the detection resistance selected for each level and the voltage range on the detection resistance.

c) Design of the secondary voltage for the transformer: The secondary voltage for the transformer is equal to the product of the total resistance (including the internal resistance of the transformer) and the largest current （3240A） in the test loop. Suppose the sum of the test sample resistance, the transformer resistance and the wire resistance is 6 mΩ, the total resistance shall be 8 mΩ when plus the current detection resistance of 2 mΩ. Thus the largest output voltage for the secondary level of the transformer will be 26V. With consideration to some design margins, it is determined to be 27V.

Table 8-2 Detection levels

Level number	1	2	3	4
Range of rated current （A）	100～225	40～100	16～40	6～16

Table 8-3 The current detection resistance and the voltage range for current detection

Level number	1	2	3	4
Current detection current range(A)	100～3240	40～1440	16～576	6～230
Current detection resistance (mΩ)	2	4	12	30
Voltage range for current detection(V)	0.2～6.5	0.16～5.8	0.19～6.9	0.18～6.9

d) Design of the balance resistance R2: There are several levels for the output voltage of the voltage regulator. To minimize the relative errors caused by step adjustments, it is hoped to operate in the high voltage area as practicable as possible. According to the voltage range for current detection in Table 8-3, and considering the voltage drops on the resistance of the test sample, the wire resistance and the internal resistance of the transformer, the secondary operating voltage for level 1 of the transformer is 0.2～6.5V. Through the switch between J4, J3, and J2, the largest value in voltage range of the corresponding primary side is 26×8=208V, the lowest value in such voltage range is 0.8×8×12=76.8V. The secondary operating voltage for level 4 of the transformer is 0.216 to 8.28V. The voltage range of the corresponding primary side is only 20.7 to 66.24V. The distance between operating voltage ranges of level 1 and level 4 is too large, thus a balance resistance shall be introduced to compensate this defect. Tale 8-4 shows the voltage operating range after introducing a balance resistance.

2）Design of the computer detection and control circuit：
a) Detection circuit. The signals required to be detected in the instantaneous protecting reliability test for the circuit breaker include three

parameters, such as the power supply voltage U_y, the output voltage of the strong current transformer U_f, and the rated current passing the test sample I_N. In Figure 8-19, three important parameters required to be detected are indicated, among which refers to the shunt set for the detection of the loop current.

Table 8-4 Operating voltage range

Level number	1	2	3	4
Current detection range（A）	100~3240	40~1440	16~576	6~230
Current detection resistance (mΩ)	2	4	12	30
Balance resistance (mΩ)	0	7	25	76
Total resistance (mΩ)	8	17	43	112
Secondary operating voltage range (V）	0.8~26	0.7~24.5	0.7~24	0.7~25.8
Output voltage range of voltage regulator (V）	77~208	67~196	67~192	67~206

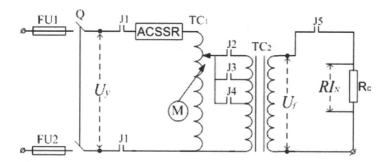

Figure 8-19 Sketch map for the detecting locations of parameters

These detection signals should be modulated to be signals within the A/D range after isolation process. The power supply signal U_y can use the isolation transformer with the voltage ratio of 62, thus the largest peak value corresponding to 250V（AC） is $\pm 250\sqrt{2} \div 62 = \pm 6\text{V}$, which should be limited within the A/D range. The signals of the output voltage of the strong current transformer U_f and the loop current I_N will be input into the computer through the isolation module. The A/D board of the computer employs Advantech PCL818HG, with single ended working mode and the range of $\pm 10\text{V}$. The sketch map for signal acquisition is as shown in Figure 8-20.

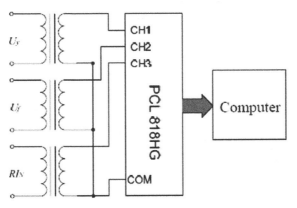

Figure 8-20 Sketch map for the signal acquisition

b) The control circuit. Figure 8-21 shows the schematic diagram for the control part. K1 and K2 is employed to control the positive and negative rotation of the motor for the power driven voltage regulator, both of which should be interlocked. The computer controls the AC solid state relay ACSSR through I/O board to achieve the electrifying control of field coils of the relays K1 and K2 and the contactor from J1 to J5. Besides, it shall control the phase selection switches through the solid state relays to achieve the selection of phases and switching on.

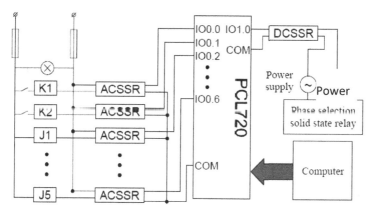

Figure 8-21 Schematic diagram for the control circuit

When performing the instantaneous characteristic test for the circuit breaker, the test current is achieved through the control of the output voltage of the power driven voltage regulator. The power driven voltage regulator comprises of the voltage regulator, and the decelerator and the servo motor which is in coaxial junction with the regulator.

5.3.5 DESIGN OF THE SOFTWARE

Design of the software for the instantaneous protecting reliability test of the circuit breaker mainly comprises the following technical problems:
(1) To achieve the control timing sequence of the molded case circuit breaker and the acquisition of the test parameters, the computer is required to fulfill timing operations, i.e. the timing interruption technology of the computer.
(2) To facilitate users to monitor the whole process and the operation of users, the software system should be designed to be the graphics mode, which is competent to drawing the dynastic curve for the test parameters in real time.

Major functions of the software are as follows:
(1) Test parameters setup: It is used to receive parameters such as the test sample name, the rated current, the multiple of the upper limit current, the multiple of the lower limit current and the limit of release time and so on.
(2) Control the test: Control the generation of the current, the lower limit current and the upper limit current specified by users; Gather signals of the

voltage and current; Judge the release situations of the test sample; Calculate the power factor angle; Achieve the selection of phases and switching on. Employ timing operation technology. Control statuses of each contactor and the test voltage according to the time requirements for each component of the control timing sequences. Check whether a failure occurs. Save the failure data when it happens. Display test parameters and the parameter dynastic curves in real time when accomplishing the control.

(3) Display and print of the test data: Provide the function for users to check the test data.

Through above analysis of the system requirements, the software system shall comprise four levels of function modules. The calling relationship of the upper three levels of modules is as shown in Figure 8-22.

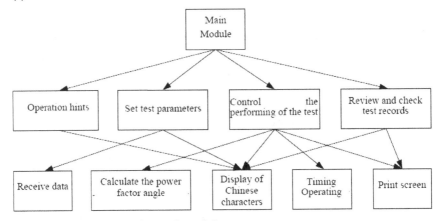

Figure 8-22 Structure chart of modules

The test current can be quickly and accurately obtained through the computer-controlled instantaneous protecting reliability test of the circuit breaker, which eliminates the aperiodic current and with much higher accuracy. With the computer digital control manner, we can conveniently analyze the accuracy of the instantaneous characteristic, and it provides accurate test data for improving the product quality.

9

RELIABILITY OF OVERLOAD RELAY

Overload relay is widely applied in motor overload protection. For many people there is concern whether the overload relay can correctly and reliably do the protective function. The protective function has two criteria: (i) no OPEN should occur while the overload relay is energized with 1.05x rated current from its cold state to hot state; and (ii) OPEN should occur in 2 seconds when the test sample is energized with 7.2x rated current.

The second criteria also means that relay OPEN should occur (a) at 1.2x rated current in 2h; (b) at 1.5x rated current in 2 minutes; (c) at 7.2 x rated current in 2 to 10 seconds from its cold state. Its reliability is mainly examined in terms of its miss operation and false operation.

To study the reliability of overload relay, we should first study the reliability indices of overload relay. It should be capable of reflecting the overall and quantitative reliability of the overload relay. Under the condition of overall reliability characteristics, the product should reflect minimum reliability indices. The reliability indices should be established after in-depth analysis of the operating characteristics and fault modes of overload relay. The reliability indices reflect the application requirements and the weaknesses of the current product.

1   RELIABILITY INDEX

Overload relay can have with two fault modes:
(1)   Did not CLOSE reliably (miss operation).
(2)   OPEN in a supposedly CLOSE situation (false operation).

Considering the operating characteristics of overload relay as a protective electrical apparatus and its two fault modes of miss operation and false operation, we adopt the level of protective success rate R (abbreviated as success rate) as its reliability index pertinent to these two fault modes. The protective success rate is referred to the probability that the product can complete the specified function under the specified conditions, or the probability of the success that the product tests under certain conditions.

The success rate grades of overload relay can be divided into five grades according to the values of unacceptable success rate R_1 (grades I, II, III, IV and V), as shown in Table 9-1.

Table 9-1 Success rate grades and R_1 values

Success Rate Grade	R_1
I	0.99
II	0.98
III	0.97
IV	0.96
V	0.95

2 TEST REQUIREMENTS

In each verification test of two reliability indices of overload relay, there are two types of test plans that can be selected. They are (i) time or failure curtailed test plan and (ii) curtailed sequential test plan. At present, the annual output of overload relay has reached to hundreds of millions sets, with economical price. Therefore, it's recommended to adopt time or failure curtailed test plan for examination in the success rate verification test of overload relay.

2.1 ENVIRONMENT CONDITIONS

The test should be conducted according to the standard atmospheric conditions specified in the Chinese National Standard GB2421-1999 *Environmental Testing for Electric and Electronic Products--Part 1: General and Guidance*:

- Temperature: 15~25°C;
- Relative humidity: 45%~75%;
- Atmospheric pressure: 86~106kPa

The test sample should be placed in the standard atmospheric conditions for enough time, not less than 8 hours, to let the test sample reach thermal equilibrium.

2.2 INSTALLATION CONDITIONS

(1) The test sample should be installed in a place of normal application.
(2) The test sample should be installed in a place without remarkable impact and vibration.
(3) The inclination between the installing surface and vertical surface of the test sample should be consistent with the product standard.

2.3 CONDITIONS OF POWER SUPPLY

Adopt a constant current source that is a sine wave power supply with a frequency of 50Hz and an allowable current deviation of ±5%.

2.4 CONDITIONS OF EXCITATION

In reliability test of overload relay operating protection, energize the overload relay with 1.05x rating current from cold state to hot state, and then increase the rated current to 1.2x or 1.5x; or energize it with 7.2x rated current from its cold state.

3   TEST METHOD

3.1 TEST ITEMS

(1) Normal operating current test: the test sample should not OPEN while energized with 1.05X rating current from its cold state for 2h.
(2) 1.2X overload current test: the test sample should OPEN within the time as specified in the product standard while energized with 1.2X rated current after heating to stable state.
(3) 1.5X overload current test: the test sample should OPEN within the time as specified in the product standard while energized with 1.5X rated current after heating to stable state.

(4) 7.2X overload current test: the test sample should OPEN within the disengagement time as specified in the standard while energized with 7.2x rating current from its cold state.

The times of these four tests should take up respectively one fourth of the reliability test cutoff time n_2.

3.2 PREPARATION OF TEST SAMPLES

The test samples should be randomly sampled from the conforming products which are produced by mass production under stable process conditions. The quantity of products for sampling should not be less than 10X of the test sample quantity n.

3.3 INSPECTION OF TEST SAMPLES

(1)   Inspection before the test

To check whether the components of test samples are damaged or broken due to transportation, To eliminate damaged test samples and replenish new ones according to regulations; the eliminated test samples wouldn't be included in the accumulated failure sample number r

(2)   Inspection during the test

In the test, when energizing 1.05x rating current from cold state to hot state, observe whether the test sample operates within 2h, and then raise the rated current to 1.2X or 1.5X, or energize 7.2X rated current from cold state, to observe whether the product operates within the specified disengagement time range. The test sample should reset after each action, and in case the product comes with both manual and automatic reset modes, theX of each reset mode should take up half of total reset times.

3.4 FAILURE CRITERIA

(1) When energized with 1.05X rated current from cold state to hot state, the test sample has operated, or when energized with 7.2X rated current from cold state, the test sample has operated with 2s (i.e. occur false operation); when energized with 1.2X rated current, the test sample fails function within 2h, or when energized with 1.5X rated current, the test sample fails function within

2*min*, or when energized with 7.2X rated current, the test sample fails function within the specified disengagement time range (i.e. occur miss operation).

(2) After the operation of test sample, the normally on contact cannot be reliably closed or the normally off contact cannot reliably open.

4 TEST PLAN AND PROCEDURE OF RELIABILITY VERIFICATION TEST

4.1 SUCCESS RATE VERIFICATION TEST PLAN

Success ratio refers to the probability of products to perform the intended function for a specified interval of time under specified conditions. It is also referred to as the probability of test success under specified conditions. In terms of overload relay, it refers to the probability that the overload relay miss-operation or false operation does not occur during its operating period.

Success rate verification test plan under the conditions of two parameters R_1 and β is as shown in Table 9-2.

Table 9-2 Success rate verification test plan ($\beta = 0.1$)

R_1 \ A_C \ n_f	1	2	3	4	5	6	7	8	9
0.95	77	105	132	158	184	209	234	258	282
0.96	96	132	166	198	230	262	292	323	333
0.97	129	176	221	265	308	349	390	431	471
0.98	194	265	333	398	462	525	587	648	708
0.99	388	531	667	798	926	1051	1175	1297	1418

4.2 TEST PROCEDURE

The test should be conducted according to the following procedure:

(1) To select product success rate index (success rate grade).

(2) To select A_C.

(3) According the selected reliability grade and A_C, find in Table 9-2 the test number n_f required by acceptance judgment.

(4) To select test stopping times n_z of the test sample, and it's recommended to select n_z between 40~80 times.

(5) According to n_f, n_z and A_C, determine the test sample number n by equation (9-1).

$$n = \frac{n_f}{n_z} + A_C \qquad\qquad (9\text{-}1)$$

(6) Randomly sample n test samples from the conforming products that are manufactured by lot and screened.

(7) Conduct test and inspection, and stop testing the test sample when one test sample is failure.

(8) To statistic the total failure times of all test samples r_Σ ($r_\Sigma = r_1 + r_2$); wherein r_1 is miss operation time and r_2 is false operation time.

(9) Test result determination:

1) The test could be judged to be qualified (acceptance) when the accumulative test number n_Σ is no less than the test number n_f which is required for making accepetance decision whilst the whole failure number r_Σ is no more than truncated failure number r_C ($r_\Sigma \leq A_C$).

2) The test could be judged to be unqualified(rejection) when the accumulative test number n_Σ is less than the test number n_f which is required for making accepetance judge whilst the whole failure number r_Σ is more than truncated failure number r_C ($r_\Sigma \leq A_C$)

5   RELIABILITY TEST DEVICE

Advanced and complete reliability test device is the basis for product reliability verification test. To improve the efficiency and correctness of reliability verification test, the reliability test device should be developed according to the product failure criteria and reliability verification test method.

5.1 TECHNICAL PERFORMANCES OF RELIABILITY TEST DEVICE

(1) It should be capable of performing reliability verification test of various overload relays, such as thermal relay and electronic overload protective relay.

Test control cabinet   Constant current source cabinet

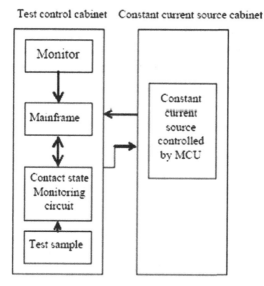

Figure 9-1 Principle block diagram of reliability test device of overload relay

(2) According to test requirement, the test personnel can set and calibrate all reliability verification test related test parameters by the test device software, such as the test time under 1.05X rated current, and the test time under 1.2X rated current, etc.

(3) During reliability test, it's capable of operating time detection under 1.2, 1.5, and 7.2 times rated current. The test under 7.2 times rated current starts directly from cold state. The operating time detection under 1.2 and 1.5 times rated current: detect the operating time after energizing 1.05 times rated current from cold state to hot state, and judge whether false operation will occur from cold state to hot state.

(4) The current of test device is output from constant current source, with maximum output current of 300A and minimum current of 10A. The current is stable, highly precise, with rapid respond time to interference. The test current can be manually fine adjustment.

(5) Test device can automatically record the test times and test sample operating time, and judge the situations of test sample false operation and miss operation, and inquiry and print the test data.

(6) The device has complete date protection performances. It can avoid loss of data after accidental power loss, and cannot damage to the acquired data after power recovery.

(7) The Device can test for both single test sample and 3 test samples.

(8) During the test, the device is capable of operating prompts, reducing incorrect operation and stopping the test at any time, with easy operation and good human-machine interface.

5.2 HARDWARE DESIGN OF RELIABILITY TEST DEVICE

The reliability test device of overload relay is mainly composed of computer test control cabinet and constant current source cabinet, with principle block diagram as shown in Figure 9-1

5.2.1 TEST CONTROL CABINET

Test control cabinet is composed of industrial PC, contact state monitoring circuit, current converter and control circuit, and test sample connecting circuit, etc. Industrial PC comes with the same functions, application method and software development environment of commercial PC. However, the positive pressure wind cooling of industrial PC is designed to effectively prevent over temperature. Industrial PC comes with the sealing door of floppy disc drive and air filter, effectively preventing dust from entering. The damping pressure bar and buffer bracket protects the industrial PC against impact and vibration. The full steel sealed enclosure can prevent electromagnetic interference. Different from general commercial PC that comes with the system motherboard structure of large board, industrial PC adopts modular structure, with passive motherboard, CPU in place of system board, and several expansion slots, for easy hardware development. Considering these characteristics of industrial PC, we adopt industrial PC as the master computer of reliability test device of overload relay.

Contact state monitoring circuit is mainly to judge the state of overload relay contact, i.e. to detect the operating situations of overload relay under testing current. This circuit sends signal to industrial control computer after photoelectric isolation.

Current converter and control circuit is mainly composed of 8 contactors and 8 solid-state relays, within which, 2 contactors are used to switch constant current source circuit, and the other 6 contactors are used to switch the 3 est sample circuits; the test of one test sample is controlled by 2 contactors, so not only single test sample can be tested, but also 3 test samples can be tested at the same time.

The test device comes with 6 wiring terminals, 1 test sample can be connected between each pair of upper and lower terminals, and the poles of overload relay should be connected in series.

5.2.2   CONSTANT CURRENT SOURCE CABINET

Constant current source cabinet is a power supply cabinet controlled independently by MCU (Micro-Control Unit), with 10A~300A output current and 2.7KW power. The current source comes with two output loops of 50A and 300A. The maximum output voltage of 50A output loop is up to 24V, and the maximum voltage of 300A output loop is up to 9V.

5.3 SOFTWARE DESIGN OF TEST DEVICE

The application software for overload relay reliability verification test is compiled by QB language, as a high-level language, it provides the interface command for connecting to hardware, so it's capable of both direct hardware control and easy to write more complicated human-machine interface program.

The control software is designed in modular mode, i.e. each function is completed by some relatively independent function modules. Thus, it does not only convenient to respectively use and software debugging, but also strengthen the reliability of test running. The software comes with prompts pertinent to operations, and excellent human-machine dialogue interface, easy for the test personnel to operate the device.

The test software mainly includes four key modules of test running, data display, data access, and test parameters modification. Data display function is mainly for the user to easily read the test results at any time, to learn about the test situation of each test sample. Data access module is capable of saving the test result as other file, for long-term storage of test results by the user. In new test, it's also required to save the original test data otherwise. The original data files can also be loaded in the computer for user to inquiry or print. The function of test parameters modification modules is mainly used for the user to modify the main test parameters, and the test parameters of different test samples may be different, this module provides the function of test parameters modification for users.

Test running module, as the key part of the software, is responsible for controlling the automatic operation of test device and storage of test data throughout the test. Test running mainly includes three tests, i.e. overload relay reliability test under 1.2 times rated current, overload relay reliability test under 1.5 times rated current, and overload relay reliability test under 7.2 times rated current.

In the reliability tests of overload relay under 1.2 and 1.5 times rated current, first all test samples are energized 1.05 times rated current from cold state to hot state, and then energized the corresponding testing current. In the tests, the computer automatically monitor whether the overload relay operates or not. In case the test sample does not operate while energized with 1.05 times rated current from cold state to hot state, the final "state before overload" should be recorded as "hot state"; after the heating time reaches the specified time, conduct the reliability tests of overload relay under 1.2 and 1.5 times rated current, and record the action states of test sample during the test; in case the test sample does not operate within the specified time, the test time should be recorded as negative value, this indicates that the test sample does not operate; in case of the test sample false operation while it's energized with 1.05 times rated current from cold state to hot state, the final "state before overload" should be recorded as "function", the test should be automatically exited, and the overload test will be conducted no more.

The reliability rest of overload relay under 7.2 times rated current is started from its cold state. The operating time of overload relay under 7.2 times rated current is considered a time range, with its upper and lower limits. During the test, the computer automatically monitor whether the overload relay operates or not. The action states of test sample should be recorded during the test, in case the test sample does not operate within the specified upper time limit, the test time should be recorded as negative value, this indicates that the test sample does not operate; in case the test sample false operation within the specified lower time limit, the final testing "state before overload" should be recorded as "Action"; the final testing "state before overload" should be recorded as "cold state" in case the operating time exceeds the specified lower time limit.   During the test, the test can be dropped out or skipped over to the next test at any time.

REFERENCE

Lu Jianguo and Wang Jingqin. *Reliability Theory and Its Application of Electrical Apparatus. China Machine Press* (in Chinese), 1996

Lu Jianguo, *Reliability of Electric Products.* China Machine Press (in Chinese), 1991

Dai Shusen, Fei Heliang, Wang Lingling, Su Deqing, Bai Hexiang and Teng Huailiu. *Reliability Test and Its Statistical Analysis.* China Machine Press (in Chinese), 1983

Xiao Dehui, *Reliability Engineering.* China Aerospace Press (in Chinese), 1985

Mao Shisong and Wang Lingling, *Reliability Statistics.* East China Normal University Press (in Chinese), 1984

Guo Yongji, *Reliability Principle of Power System and Its Application*, China Tsinghua University Press (in Chinese), 1983

Mathematics Department of Chinese School of Mines. Mathematics Handbook. Science Press (in Chinese), 1981

China National Standard: GB/T 2689.2-1981 Figures Estimation Method of Life Test and Accelerated Life Test (for Weibull distribution). Standards Press of China (in Chinese)

Mao Shisong and Luo Chaobin, Reliability Analysis of No Failure Data. Journal of Mathematical Statistics and Applied Probability (in Chinese), 1989 (4):489 ~507

Zhang Zhihua, Statistical Analysis of No Failure Data, Journal of Mathematical Statistics and Applied Probability (in Chinese), 1995 (10):94~101

Zhang Zhongzhan and Yang Zhenhai, Statistical Analysis of No Failure Data, Journal of Mathematical Statistics and Applied Probability (in Chinese). 1989 (4):508~515

Zhang Jichang. Bayes Analysis of No Failure Data. Higher Applied Mathematics Xuebao (in Chinese)，1995 (10):19~ 2

China National Standard: GB/T 5080.5-1985 Reliability test of equipment. Success Rate Verification Test Plan. Standards Press of China (in Chinese)

Qian Neng, *Programming Tutorial of C++*. China Tsinghua University Press (in Chinese)，2001

Lu Jianguo and Su Xiuping. *Theory and Application of Reliability Optimization Design of Electromagnetic System on Electrical Apparatus*, China Machine Press (in Chinese)，2003

Dong Yuge, *Fuzzy Reliability Design on Machine*, China Machine Press (in Chinese)，2000

Lu Jianguo and Jin Fuqun. Reliability Design on Electromagnetic System of Relay. Oklahoma: Proceedings of 36th Relay Conference, 1988, 71~ 74

Jin Fuqun and Lu Jianguo. Reliability Mathematical Model for Combined Reserve Design Circuit and Its Application in Relay. Oklahoma: Proceedings of 38th Relay Conference, 1990

Lu Jianguo, Li Zhigang and Wang Jingqin. The Device of Research on Relay Contact Reliability. Montreal: Proceedings of 36th IEEE Holm Conference on Electrical Contacts and 15th International Conference on Electrical Contacts, 1990, 398~ 402

Hu Changshou, et al. *Relaibility Design Handbook for Aerospace*. China Machine Press (in Chinese)，1999.

Mou Zhizhong, *Relaibility Design*. China Machine Press (in Chinese)，1993

Lu Yuming, *Reliability Design on Machine Part*. Higher Education Press (in Chinese)，1989

Su Xiuping, Study on the Reliability Design Techniques and Optimization Design Techniques of Electromagnetic Mechanism in Electrical Apparatus. Dissertation for Ph.D. Hebei University of Technology (in Chinese)，1999

Fei Hongjun and Zhang Guansheng. *Dynamic Analysis and Calculate for Electromagnetic System*. China Machine Press (in Chinese)，1993

Su Xiuping and Lu Jianguo. Study on the Reliability Optimization Design of Electromagnetic System in Miniature Relays. Proceedings of 47th Annual International Relay Conference, 1999, 181~187

Su Xiuping and Lu Jianguo. The Study on the Reliability Optimization Design of Electromagnetic System in Electrical Apparatus. Proceedings of 4th International Conference on Reliability, Maintainability & Safety, 1999, 756~761

Su Xiuping, Lu Jianguo, et al. The Study on the Theory for the Pick-up Reliability of Electrical Apparatus. Proceedings of 3rd Electrical Contact, Arcs, Apparatus and their Applications, 1997, 614~617

Su Xiuping, Lu Jianguo, Liu Guojin, Bai Huizhen, Li Junqing. Multi-bjective Optimization of AC Electromagnetic System based on Game Theory. The Fourth International Conference on Electromagnetic Field Problems & Applications, 2000, 234~237

Chen, G.B., *Cryogenic Insulation and Heat Transfer*. Zhejiang University Press (in Chinese), 1989, 240~256

Chen. X.D., Zhu. Y.X., Wang. Y.Q., Qu. S., Gu, H., Reliability test research of molded case circuit breaker and its application. Low Voltage Apparatus (in Chinese)., 2001. (2):8~10

Hanna. G.L., Retrofit breakers resolve trip failures. Power Engineering, 1995. 99(3):38~40.

National Standard GB 14048.2-2001, Low Voltage Switchgear and Control Equipment: Low Voltage Circuit Breakers. Chinese Standards Press, Beijing (in Chinese), 2002

Norris, A., Report of Circuit Breaker Reliability Survey of Industrial and Commercial Installations. IEEE Conference Record of Industrial and Commercial Power Systems Technical Conference. Chicago, IL, USA. 1989

O'Donnell. P., Braun, W.F., Heising. C.R., Khera. P.P., Kornblit. M. and McDonald. K.D., 1997. Survey results of low-voltage circuit breakers as found during maintenance testing: working group report. IEEE Transactions on Industry Applications. 33(5): 1367~1369

Rieder, W.F., Strof, T.W., Relay Life Tests with Contact Resistance Measurement after Each Operation. Proceedings of the 36th IEEE Holm Conference on Electrical Contacts & the 15th International Conference on Electrical Contacts. International Conference and Tour Montreal Quebec. Canada, 1990.73~78

Schoen, P.E., Circuit breaker testing technology (Part 1). Electrical Manufacturing. 1989a.3(5):37~39

Schoen. P.E., Circuit breaker testing technology (Part 2). Electrical Manufacturing. 1989b. 3(6):21~24

APPENDIX 1 Γ FUNCTION TABLE

m	$\Gamma(\dfrac{1}{m}+1)$	m	$\Gamma(\dfrac{1}{m}+1)$	m	$\Gamma(\dfrac{1}{m}+1)$
0.1	11!	1.5	0.903	2.8	0.890
0.2	6!	1.6	0.897	2.9	0.892
0.3	9.260	1.7	0.892	3.0	0.894
0.4	3.323	1.8	0.889	3.1	0.894
0.5	2.000	1.9	0.887	3.2	0.896
0.6	1.505	2.0	0.886	3.3	0.897
0.7	1.266	2.1	0.886	3.4	0.898
0.8	1.133	2.2	0.886	3.5	0.900
0.9	1.052	2.3	0.886	3.6	0.901
1.0	1.000	2.4	0.886	3.7	0.902
1.1	0.965	2.5	0.887	3.8	0.904
1.2	0.941	2.6	0.888	3.9	0.905
1.3	0.923	2.7	0.889	4.0	0.906
1.4	0.911				

APPENDIX 2 - UNDER SIDE QUANTILE $\chi_p^2(f)$
OF X2 DISTRIBUTION

$\chi_p^2(f)$ $\diagdown$ P $\diagdown$ f	0.01	0.025	0.05	0.10	0.25	0.75	0.90	0.95	0.975	0.99
1		0.001	0.004	0.016	0.10	1.32	2.71	3.84	5.02	6.64
2	0.02	0.051	0.10	0.21	0.58	2.77	4.61	5.99	7.38	9.21
3	0.12	0.216	0.35	0.58	1.21	4.11	6.25	7.82	9.35	11.34
4	0.30	0.48	0.71	1.06	1.92	5.39	7.78	9.49	11.14	13.28
5	0.55	0.83	1.15	1.61	2.68	6.63	9.24	11.07	12.83	15.09
6	0.87	1.24	1.64	2.20	3.46	7.84	10.65	12.59	14.45	16.81
7	1.24	1.69	2.17	2.83	4.26	9.04	12.02	14.07	16.01	18.48
8	1.65	2.18	2.73	3.49	5.07	10.22	13.36	15.51	17.54	20.09
9	2.10	2.70	3.33	4.17	5.90	11.39	14.68	16.92	19.02	21.67
10	2.56	3.25	3.94	4.87	6.74	12.55	15.99	18.31	20.48	23.21
11	3.05	3.82	4.58	5.58	7.58	13.70	17.28	19.68	21.92	24.73
12	3.57	4.40	5.23	6.30	8.44	14.85	18.55	21.03	23.34	26.22
13	4.11	5.01	5.89	7.04	9.30	15.98	19.81	22.36	24.74	27.69
14	4.66	5.63	6.57	7.79	10.17	17.12	21.06	23.69	26.12	29.14
15	5.23	6.26	7.26	8.55	11.04	18.25	22.31	25.00	27.49	30.58
16	5.81	6.91	7.96	9.31	11.92	19.37	23.54	26.30	28.85	32.00
17	6.41	7.56	8.67	10.09	12.79	20.49	24.77	27.59	30.19	33.41
18	7.02	8.23	9.39	10.87	13.68	21.61	25.99	28.87	31.53	34.81
19	7.63	8.91	10.12	11.65	14.56	22.72	27.20	30.14	32.85	36.19
20	8.26	9.59	10.85	12.44	15.45	23.83	28.41	31.41	34.17	37.57
21	8.90	10.28	11.59	13.24	16.34	24.94	29.62	32.67	35.48	38.93
22	9.54	10.98	12.34	14.04	17.24	26.04	30.81	33.92	36.78	40.29
23	10.20	11.69	13.09	14.85	18.14	27.14	32.01	35.17	38.08	41.64
24	10.86	12.40	13.85	15.66	19.04	28.24	33.20	36.42	39.36	42.98
25	11.52	13.12	14.61	16.47	19.94	29.34	34.38	37.65	40.65	44.31
26	12.20	13.84	15.38	17.29	20.84	30.44	35.56	38.89	41.92	45.64
27	12.88	14.57	16.15	18.11	21.79	31.53	36.74	40.11	43.19	46.96
28	13.57	15.31	16.93	18.94	22.66	32.62	37.92	41.34	44.46	48.28
29	14.26	16.05	17.71	19.77	23.57	33.71	39.09	42.56	45.72	49.59
30	14.59	16.79	18.49	20.60	24.48	34.80	40.26	43.77	46.98	50.89
31	15.66	17.54	19.28	21.43	25.39	35.89	41.42	44.99	48.23	52.19
32	16.36	18.29	20.07	22.27	26.30	36.97	42.59	46.19	49.48	53.49
33	17.07	19.05	20.87	23.11	27.22	38.06	43.75	47.40	50.72	54.78
34	17.79	19.81	21.66	23.95	28.14	39.14	44.90	48.60	51.97	56.06
35	18.51	20.57	22.47	24.80	29.05	40.22	46.06	49.80	53.20	57.34
36	19.23	21.34	23.227	25.64	29.97	41.30	47.21	51.00	54.44	58.62
37	19.96	22.11	24.08	26.49	30.89	42.38	48.36	52.19	55.67	59.89

cont'd. Appendix 2

$\chi_p^2(f)$ P f	0.01	0.025	0.05	0.10	0.25	0.75	0.90	0.95	.0975	0.99
38	20.69	22.88	24.88	27.34	31.82	43.46	49.51	53.38	56.90	61.16
39	21.43	23.65	25.70	28.20	32.74	44.54	50.66	54.57	58.12	62.43
40	22.16	24.43	26.51	29.05	33.66	45.62	51.81	55.76	59.34	63.
41	22.91	25.22	27.33	29.91	34.59	46.69	52.95	56.94	60.56	64.95
42	23.65	26.00	28.14	30.77	35.51	47.77	54.09	58.12	61.78	66.21
43	24.40	26.79	28.97	31.63	36.44	48.84	55.23	59.30	62.99	67.46
44	25.15	27.58	29.79	32.49	37.36	49.91	56.37	60.48	64.20	68.71
45	25.90	28.37	30.61	33.35	38.29	50.99	57.51	61.66	65.41	69.96
46	26.66	29.16	31.44	34.22	39.22	52.06	58.64	62.83	66.62	71.20
47	27.42	29.96	32.27	35.08	40.15	53.13	59.77	64.00	67.82	72.44
48	28.18	30.76	33.10	35.95	41.08	54.20	60.91	65.17	69.02	73.68
49	28.94	31.56	33.93	36.82	42.01	55.27	62.04	66.34	70.22	74.92
50	29.71	32.36	34.76	37.69	42.94	56.33	63.17	67.51	71.42	76.15
51	30.48	33.16	35.60	38.56	43.87	57.40	64.30	68.67	72.62	77.39
52	31.25	33.97	36.44	39.43	44.81	58.47	65.42	69.83	73.81	78.62
53	32.02	34.78	37.28	40.31	45.74	59.53	66.55	70.99	75.00	79.84
54	32.79	35.59	38.12	41.18	46.68	60.60	67.67	72.15	76.19	81.07
55	33.57	36.40	38.96	42.06	47.61	61.67	68.80	73.31	77.38	82.29
56	34.35	37.21	39.80	42.94	48.55	62.73	69.92	74.47	78.57	83.51
57	35.13	38.03	40.65	43.82	49.48	63.79	71.04	75.62	79.75	84.73
58	35.91	38.84	41.49	44.70	50.42	64.86	72.16	76.78	80.94	85.95
59	36.70	39.66	42.34	45.58	51.36	65.92	73.28	77.03	82.12	87.17
60	37.49	40.48	43.19	46.46	52.29	66.98	74.40	79.08	83.30	88.38
61	38.27	41.30	44.04	47.34	53.23	68.04	75.51	80.23	84.47	89.59
62	39.06	42.13	44.89	48.23	54.17	69.10	76.63	81.38	85.65	90.80
63	39.86	42.95	45.74	49.11	55.11	70.17	77.75	82.53	86.83	92.01
64	40.65	43.78	46.60	50.00	56.05	71.23	78.86	83.68	88.00	93.22
65	41.44	44.60	47.45	50.88	56.99	72.29	79.97	84.82	89.18	94.42
66	42.24	45.43	48.31	51.77	57.93	73.34	81.09	85.97	90.35	95.63
67	43.04	46.26	49.16	52.66	58.87	74.40	82.20	87.11	91.52	96.83
68	43.84	47.09	50.02	53.55	59.81	75.46	83.31	88.25	92.69	98.03
69	44.64	47.92	50.88	54.44	60.76	76.52	84.42	89.39	93.86	99.23
70	45.44	48.76	51.74	55.33	61.70	77.58	85.53	90.53	95.02	100.43
71	46.25	49.59	52.60	56.22	62.64	78.63	86.64	91.67	96.19	101.62
72	47.05	50.43	53.46	57.11	63.59	79.69	87.74	92.81	97.35	102.82
73	47.86	51.27	54.33	58.01	64.53	80.75	88.85	93.95	98.52	104.01
74	48.67	52.10	55.19	58.90	65.47	81.80	89.96	95.08	99.68	105.20
75	49.48	52.94	56.05	59.80	66.42	82.86	91.06	96.22	100.84	106.39

cont'd. Appendix 2

$\chi^2_p(f)$	0.01	0.025	0.05	0.10	0.25	0.75	0.90	0.95	.0975	0.99
76	50.29	53.78	56.92	60.69	67.36	83.91	92.17	97.35	102.00	107.58
77	51.10	54.62	57.79	61.59	68.31	84.97	93.27	98.48	103.16	108.77
78	51.91	55.47	58.65	62.48	69.25	86.02	94.37	99.62	104.32	109.96
79	52.73	56.31	59.52	63.38	70.20	87.08	95.48	100.75	105.47	111.14
80	53.54	57.15	60.39	64.28	71.15	88.13	96.58	101.88	106.63	112.33
81	54.36	58.00	61.26	65.18	72.09	89.18	97.68	103.01	107.78	113.51
82	55.17	58.85	62.13	66.08	73.04	90.24	98.78	104.14	108.94	114.70
83	55.99	59.69	63.00	66.69	73.99	91.29	99.88	105.27	110.09	115.88
84	56.81	60.54	63.88	67.88	74.93	92.34	100.98	106.40	111.24	117.06
85	57.63	61.39	64.75	68.78	75.88	93.39	102.08	107.52	112.39	118.24
86	58.46	62.24	65.62	69.68	76.83	99.45	103.18	108.65	113.54	119.41
87	59.28	63.10	66.50	70.58	77.72	95.50	104.28	109.77	114.69	120.59
88	60.10	63.94	67.37	71.48	78.73	96.55	105.37	110.90	115.84	121.77
89	60.93	64.79	68.25	72.39	79.68	97.60	106.47	112.02	116.99	122.94
90	61.75	65.65	69.13	73.29	80.63	98.65	107.57	113.15	118.14	124.12
91	62.58	66.50	70.00	74.20	81.57	99.70	108.66	114.27	119.28	125.29
92	63.41	67.36	70.88	75.10	82.52	100.75	109.76	115.39	120.43	126.46
93	64.24	68.21	71.76	76.01	83.47	101.80	110.85	116.51	121.57	127.63
94	65.07	69.07	72.64	76.91	84.43	102.85	111.94	117.63	122.72	128.80
95	65.90	69.93	73.52	77.82	85.38	103.90	113.04	118.75	123.86	129.97
96	66.73	70.78	74.40	78.73	86.33	104.95	114.13	119.87	125.00	131.14
97	67.56	71.64	75.28	79.63	87.28	106.00	115.22	120.99	126.14	132.31
98	68.40	72.50	76.16	80.54	88.23	107.05	116.32	122.11	127.28	133.48
99	69.23	73.36	77.05	81.45	89.18	108.09	117.41	123.23	128.42	134.64
100	70.07	74.22	77.93	82.36	90.13	109.14	118.50	124.34	129.56	135.81

APPENDIX 3 NUMERICAL TABLES OF $\Phi(z) = \dfrac{1}{\sqrt{2\pi}} \displaystyle\int_{-\infty}^{z} e^{-v^2/2} \, dv$

z	-0.00	-0.01	-0.02	-0.03	-0.04	-0.05	-0.06	-0.07	-0.08	-0.09
-3.0	0.00135	0.00131	0.00122	0.00126	0.00118	0.00114	0.00111	0.00107	0.00104	0.00100
-2.9	0.00187	0.00181	0.00175	0.00170	0.00164	0.00159	0.00154	0.00149	0.00144	0.00140
-2.8	0.00256	0.00248	0.00240	0.00233	0.00226	0.00219	0.00212	0.00205	0.00199	0.00193
-2.7	0.00347	0.00336	0.00326	0.00317	0.00307	0.00298	0.00289	0.00280	0.00272	0.00264
-2.6	0.00466	0.00453	0.00440	0.00427	0.00415	0.00403	0.00391	0.00379	0.00368	0.00357
-2.5	0.00621	0.00604	0.00587	0.00570	0.00554	0.00539	0.00523	0.00509	0.00494	0.00480
-2.4	0.0082	0.0080	0.0078	0.0076	0.0073	0.0071	0.0070	0.0068	0.0066	0.0064
-2.3	0.0107	0.0104	0.0102	0.0090	0.0096	0.0094	0.0091	0.0089	0.0087	0.0084
-2.2	0.0139	0.0136	0.0132	0.0129	0.0126	0.0122	0.0119	0.0116	0.0133	0.0110
-2.1	0.0179	0.0174	0.0710	0.0166	0.0162	0.0158	0.0154	0.0150	0.0146	0.0143
-2.0	0.0228	0.0222	0.0217	0.0212	0.0207	0.0202	0.0197	0.0192	0.0188	0.0183
-1.9	0.0287	0.0281	0.0274	0.0268	0.0262	0.0256	0.0250	0.0244	0.0238	0.0233
-1.8	0.0359	0.0352	0.0344	0.0336	0.0329	0.0322	0.0314	0.0307	0.0300	0.0294
-1.7	0.0446	0.0436	0.0427	0.0418	0.0409	0.0401	0.0392	0.0384	0.0375	0.0367
-1.6	0.0548	0.0537	0.0562	0.0516	0.0505	0.0495	0.0485	0.0475	0.0465	0.0455
-1.5	0.0668	0.0655	0.0643	0.0630	0.0618	0.0606	0.0594	0.0582	0.0570	0.0559
-1.4	0.0808	0.0793	0.0778	0.0764	0.0749	0.0735	0.0722	0.0708	0.0694	0.0681
-1.3	0.0968	0.0951	0.0934	0.0918	0.0901	0.0885	0.0869	0.0853	0.0838	0.0823
-1.2	0.1151	0.1131	0.1112	0.1093	0.1075	0.1056	0.1038	0.1020	0.1003	0.0985
-1.1	0.1357	0.1335	0.1314	0.1292	0.1271	0.1251	0.1230	0.1210	0.1190	0.1170
-1.0	0.1587	0.1562	0.1539	0.1515	0.1492	0.1469	0.1446	0.1423	0.1401	0.1379
-0.9	0.1841	0.1814	0.1788	0.1762	0.1736	0.1711	0.1685	0.1660	0.1635	0.1611
-0.8	0.2119	0.2090	0.2061	0.2033	0.2005	0.1977	0.1949	0.1922	0.1894	0.1867
-0.7	0.2420	0.2389	0.2358	0.2327	0.2297	0.2266	0.2236	0.2206	0.2177	0.2148
-0.6	0.2743	0.2709	0.2676	0.2643	0.2611	0.2578	0.2546	0.2514	0.2483	0.2451
-0.5	0.3085	0.3050	0.3015	0.2981	0.2946	0.2912	0.2877	0.2843	0.2810	0.2776
-0.4	0.3446	0.3409	0.3372	0.3336	0.3300	0.3264	0.3228	0.3192	0.3156	0.3121
-0.3	0.3821	0.3783	0.3745	0.3707	0.3669	0.3632	0.3594	0.3557	0.3520	0.3483
-0.2	0.4207	0.4168	0.4129	0.4090	0.4052	0.4013	0.3974	0.3936	0.3897	0.3859
-0.1	0.4602	0.4562	0.4522	0.4483	0.4443	0.4404	0.4364	0.4325	0.4286	0.4247
-0.0	0.5000	0.4960	0.4920	0.4880	0.4840	0.4801	0.4761	0.4721	0.4681	0.4641

Cont'd Appendix 3

z	0.00	0.01	0.02	0.03	0.04	0.05	0.06	0.07	0.08	0.09
0.0	0.5000	0.5040	0.5080	0.5120	0.5160	0.5199	0.5239	0.5279	0.5319	0.5359
0.1	0.5398	0.5438	0.5478	0.5517	0.5557	0.5596	0.5636	0.5676	0.5714	0.5753
0.2	0.5793	0.5832	0.5871	0.5910	0.5948	0.5987	0.6026	0.6064	0.6103	0.6141
0.3	0.6179	0.6217	0.6255	0.6293	0.6331	0.6368	0.6406	0.6443	0.6480	0.6517
0.4	0.6554	0.6591	0.6628	0.6664	0.6700	0.6736	0.6772	0.6806	0.6844	0.6879
0.5	0.6915	0.6950	0.6985	0.7019	0.7054	0.7088	0.7123	0.7157	0.7190	0.7224
0.6	0.7257	0.7291	0.7324	0.7357	0.7389	0.7422	0.7454	0.7486	0.7517	0.7549
0.7	0.7580	0.7611	0.7642	0.7673	0.7703	0.7734	0.7764	0.7794	0.7823	0.7852
0.8	0.7881	0.7910	0.7939	0.7967	0.7995	0.8032	0.8051	0.8078	0.8106	0.8133
0.9	0.8159	0.8186	0.8212	0.8238	0.8264	0.8289	0.8315	0.8340	0.8365	0.8389
1.0	0.8413	0.8438	0.8461	0.8485	0.8508	0.8531	0.8554	0.8577	0.8599	0.8621
1.1	0.8643	0.8665	0.8686	0.8708	0.8729	0.8749	0.8770	0.8790	0.8810	0.8830
1.2	0.8849	0.8869	0.8888	0.8907	0.8925	0.8944	0.8962	0.8980	0.8997	0.9015
1.3	0.9032	0.9049	0.9066	0.9082	0.9099	0.9115	0.9131	0.9147	0.9162	0.9177
1.4	0.9192	0.9207	0.9222	0.9236	0.9251	0.9265	0.9278	0.9292	0.9306	0.9319
1.5	0.9332	0.9345	0.9357	0.9370	0.9382	0.9394	0.9406	0.9418	0.9430	0.9441
1.6	0.9452	0.9463	0.9474	0.9484	0.9495	0.9505	0.9515	0.9525	0.9535	0.9545
1.7	0.9554	0.9564	0.9573	0.9582	0.9591	0.9599	0.9608	0.9616	0.9625	0.9633
1.8	0.9641	0.9648	0.9656	0.9664	0.9671	0.9678	0.9686	0.9693	0.9700	0.9706
1.9	0.9713	0.9719	0.9726	0.9732	0.9738	0.9744	0.9750	0.9756	0.9762	0.9767
2.0	0.9772	0.9778	0.9783	0.9788	0.9793	0.9798	0.9803	0.9808	0.9812	0.9817
2.1	0.9821	0.9826	0.9830	0.9834	0.9838	0.9842	0.9846	0.9850	0.9854	0.9857
2.2	0.9861	0.9864	0.9868	0.9871	0.9874	0.9878	0.9881	0.9884	0.9887	0.9890
2.3	0.9893	0.9896	0.9898	0.9901	0.9904	0.9906	0.9909	0.9911	0.9913	0.9916
2.4	0.9918	0.9920	0.9922	0.9925	0.9927	0.9929	0.9931	0.9932	0.9934	0.9936
2.5	0.99379	0.99396	0.99413	0.99430	0.99446	0.99461	0.99477	0.99492	0.99506	0.99520
2.6	0.99534	0.99547	0.99560	0.99573	0.99586	0.99598	0.99609	0.99620	0.99632	0.99643
2.7	0.99653	0.99664	0.99674	0.99683	0.99693	0.99702	0.99711	0.99720	0.99728	0.99737
2.8	0.99745	0.99752	0.99760	0.99767	0.99774	0.99781	0.99788	0.99795	0.99801	0.99807
2.9	0.99813	0.99819	0.99825	0.99831	0.99836	0.99841	0.99846	0.99851	0.99856	0.99861
3.0	0.99865	0.99869	0.99874	0.99878	0.99882	0.99886	0.99889	0.99893	0.99897	0.99900